Technician Physical Science 1

R. G. MEADOWS

B.Sc., M.Sc., Ph.D., C.Eng., M.I.E.E., M.Inst.P., A.R.C.S.

Cassell · London

Cassell Ltd: 1 St Anne's Road,
Eastbourne, East Sussex BN21 3UN

First published 1977
Second impression 1978
Third impression 1980
Second (revised) edition 1983

Printed and bound in Great Britain by Collins, Glasgow

British Library Cataloguing in Publication Data

Meadows, R.G.
  Technician physical science 1. — 2nd rev. ed.
  —(Cassell's TEC series)
  1. Physics
  I. Title
  530′.0246   QC21.2

ISBN 0-304-30959-1

Dedication to 'number one' son James, my wife Lynn, and to the
memory of Frederic, Alban and Bill.

Contents

Technician Physical Science 1

Second (Revised) Edition

This book covers the Physical Science 1 standard unit (U80/682) of the Technician Education Council certificate and diploma programmes. The learning structure and content of the syllabus has been followed exactly and the book is written with the clarity and simplicity which proved so successful in Dr Meadows' first edition. There are numerous worked examples throughout the text and problems with answers at the end of each chapter. The text is illustrated throughout with easily understood line drawings.

Dr Meadows is an experienced teacher and author and is Head of the Department of Electronic and Communications Engineering at the Polytechnic of North London. He has written widely for technician engineering courses and his books include *Technician Engineering Science 2*, *Technician Engineering Science 3*, *Technician Electronics 2*, *Technician Electronics 3* and *Technician Electrical and Electronic Principles 3*.

Preface to First Edition

This text is written specifically for students taking the level 1 unit Physical Science 1. This is a standard unit for Technician Certificate courses of the Technician Education Council in the following areas: A1 General Engineering, A2 Electronics and Communications Engineering, A3 Electrical Engineering, A4 Plant, Process and Control Engineering, A6 Fabrication and Welding Studies and Metallurgy. The standard unit is also applicable to a range of other programmes. Further Engineering Science Units will follow for levels 2 and 3.

The text follows exactly the same order as the TEC syllabus and an identical numbering sequence for the specific learning objective headings is used. I hope that this provides a rapid aid for reference purposes.

I have enjoyed writing this book and sincerely hope that students will find it both useful and interesting. Finally I should like to express thanks to all my colleagues for many helpful discussions. Special thanks to John Morgan, Roger Driscol and Norbert Singer.

RICHARD MEADOWS

April 1977

Preface to Second Edition

The text of the first edition has been thoroughly revised to take into account the changes made in the new Technician Education Council's Physical Science 1 syllabus. It follows exactly the same order as the new syllabus and has an identical numbering sequence.

RICHARD MEADOWS

December 1982

The International System of Units (SI)

The International System of Units (SI) is used throughout this book. SI units have been adopted by the International Standards Organization (ISO) and by the International Electrotechnical Commission (IEC) and are used in science and technology throughout the world.

1   The seven SI base units and two SI supplementary units

Quantity	Name of unit	Unit symbol
length	metre	m
mass	kilogram	kg
time	second	s
electric current	ampere	A
thermodynamic temperature	kelvin	K
luminous intensity	candela	cd
amount of substance	mole	mol
plane angle	radian	rad
solid angle	steradian	sr

2   The SI derived units of important quantities

Quantity	Name of unit	Unit symbol
area	square metre	m^2
volume	cubic metre	m^3
speed, velocity	metre per second	m/s
angular velocity	radian per second	rad/s
acceleration	metre per second squared	m/s^2
angular acceleration	radian per second squared	rad/s^2
density	kilogram per cubic metre	kg/m^3
frequency	hertz	Hz
force	newton	N
pressure	pascal, newton per square metre	Pa, N/m^2
torque, moment	newton metre	Nm
energy, work	joule	J
power	watt	W
electric charge	coulomb	C
electric potential	volt	V
electric field strength	volt per metre	V/m
magnetic flux	weber	Wb
magnetic flux density	tesla	T
magnetic field strength	ampere per metre	A/m
resistance	ohm	Ω
conductance	siemens	S
capacitance	farad	F
inductance	henry	H
permittivity	farad per metre	F/m
permeability	henry per metre	H/m

The SI derived units of important quantities (continued)

Quantity	Name of unit	Unit symbol
specific heat capacity	joule per kilogram kelvin	J/(kgK), $Jkg^{-1}K^{-1}$
specific latent heat capacity	joule per kilogram	J/kg
internal energy, enthalpy	joule	J
specific enthalpy	joule per kilogram	J/kg
specific volume	cubic metre per kilogram	m^3/kg
thermal conductivity	watt per metre kelvin	W/(mK), $Wm^{-1}K^{-1}$

3   Units of practical importance retained for general use

Name	Unit symbol	Value in SI units
minute	min	$1\,min = 60\,s$
hour	h	$1\,h = 60\,min = 3600\,s$
degree	°	$1° = (\pi/180)\,rad$
litre	l	$1\,l = 10^{-3}\,m^3$
tonne	t	$1\,t = 10^3\,kg$
nautical mile	1 nautical mile $= 1852\,m$	
knot	1 knot $= 1$ nautical mile per hour	
degree Celsius	°C	$T°C = T\,K - 273.15\,K$
standard atmosphere	atm	$1\,atm = 101\,325\,N/m^2$
bar	bar	$1\,bar = 10^5\,N/m^2$

4   Names and symbols for SI prefixes

Name	Symbol	Meaning (i.e. factor by which unit is multiplied)
tera	T	$\times 10^{12}$ (a million million times)
giga	G	$\times 10^9$ (a thousand million times)
mega	M	$\times 10^6$ (a million times)
kilo	k	$\times 10^3$ (a thousand times)
hecto	h	$\times 10^2$ (a hundred times)
deca	da	$\times 10^1$ (ten times)
deci	d	$\times 10^{-1}$ (a tenth)
centi	c	$\times 10^{-2}$ (a hundredth)
milli	m	$\times 10^{-3}$ (a thousandth)
micro	μ	$\times 10^{-6}$ (a millionth)
nano	n	$\times 10^{-9}$ (a thousand millionth)
pico	p	$\times 10^{-12}$ (a million millionth)

Examples
(i) $E = 212\,GN/m^2$ means $E = 212 \times 10^9\,N/m^2$
(ii) $R = 10\,M\Omega$ means $R = 10 \times 10^6\,\Omega$
(iii) $2.5\,mm^2 = 2.5 \times (10^{-3})^2 = 2.5 \times 10^{-6}\,m^2$

(iv) $f = 50\,kHz$ means $f = 50 \times 10^3\,Hz$
(v) $T = 10\,ns$ means $T = 10 \times 10^{-9}\,s$

A Material properties and statics

1 Basic chemical processes and reactions

The expected learning outcome of this chapter is that the student should be able to describe, in simple terms, the chemical processes involved in burning and rusting as examples of chemical reactions.

1.1
Air is a mixture mainly of oxygen and nitrogen

Air is a mixture of gases, the main constituents being nitrogen and oxygen. The ratio by volume is approximately four-fifths nitrogen and one-fifth oxygen. Air also contains a very small percentage by volume of the rare gases (argon, neon, helium, krypton, xenon) and the gases of carbon dioxide, methane, nitrous oxide, hydrogen, and ozone. Air may also contain other gases as a result of pollution caused by industrial processes.

The following table gives the percentage constituents of dry clean air.

Table 1.1   Constituents by volume of air (earth's atmosphere)

Nitrogen	78.09 %
Oxygen (O_2)	20.95 %
Argon (A)	0.93 %
Carbon dioxide (CO_2)	0.03 %
Neon (Ne)	0.0018 %
Helium (He)	0.0005 %
Methane (CH_4)	0.000 15 %
Krypton (Kr)	0.000 11 %
Nitrous oxide (N_2O)	0.000 05 %
Hydrogen (H_2)	0.000 05 %
Ozone (O_3)	0.000 04 %
Xenon (Xe)	0.000 01 %

1.2
The description of how a substance, such as copper, gains mass when heated in air and how oxygen is taken from the air by the copper

When a substance—and really in this context we are thinking of elements—is heated and burns in air, it undergoes a chemical change and the resulting compound formed is normally quite different from the original element. For example, when copper foil is heated in a Bunsen flame, it becomes coated with a blackish tarnish. When zinc is heated strongly, it becomes coated with a yellow powder, which changes colour to white on cooling. When lead shot is heated in an iron spoon, it becomes coated with a yellow powder. A piece of magnesium ribbon, when heated in a Bunsen flame, gives off a brilliant white flame and undergoes a chemical change from metallic form to a white powder.

In each of the above examples, the metal, when heated, takes oxygen from the air and combines with this oxygen to form a compound. The mass of the compound so formed is greater than the original mass of the element before burning.

The latter effect can be verified experimentally using the apparatus shown in fig. 1.1. A crucible, together with its lid, containing some copper foil or magnesium ribbon is first weighed. The crucible is then heated. The lid is lifted a few times during the heating process and it is observed that the metal tends to burn more vigorously. Care is also taken to ensure that as little 'smoke' as possible escapes when the lid is lifted. When all the metal is burned, the crucible and contents are re-weighed. It is observed that an increase in mass has taken place.

The increase in mass occurs owing to the metal

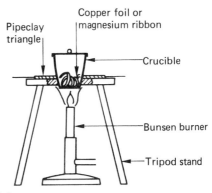

Fig. 1.1

combining with oxygen in the air to form a compound of the metal and oxygen. Note that although the experiment shows a mass increase, it does not provide conclusive proof that the gas taken from the air is oxygen. If the metal had been heated in a closed atmosphere of air, it could be observed that burning ceases when approximately one-fifth of the volume of air is used up. This, again, does not provide conclusive proof, but would strongly indicate that the metal has combined with the one-fifth oxygen contained in the air.

1.3
The description of chemical reactions as interactions between substances which involve a rearrangements of atoms

Chemical reactions occur when substances interact with each other and this interaction leads to a rearrangement of the atoms of those substances.

All substances are composed of atoms, whether as simple atoms, atoms combined to form molecules, or ions (charged atoms or parts of charged molecules). For example, mercury exists at normal temperatures as a liquid consisting of mercury atoms; oxygen and nitrogen gases are made up of molecules of O_2 and N_2; most metals consist of a crystalline lattice of positively charged metal ions, the metal atoms 'losing' one or more electrons; the compound of common salt consists of a crystalline lattice of sodium Na^+ ions and chlorine Cl^- ions; sodium hydroxide solution in water consists of Na^+ and OH^- ions.

Thus, with the general description in our minds that substances (whether elements, compounds, or mixtures) are made up of atoms, molecules, or ions, we can state that a chemical reaction occurs when two or more substances interact and a rearrangement of the atoms, molecules, or ions of these substances occurs.

1.4
The description of how substances burning in air combine with oxygen and the recognition of this as an example of a chemical reaction

Many substances, particularly metallic elements and elements such as carbon, sulphur, and phosphorus, burn in air. Some substances require heating before burning commences, but some elements, for example sodium, combine with oxygen at room temperature (a block of metallic sodium will dull if left in contact with air).

Once burning commences, it is often self-sustaining. For example, once carbon, say, in the form of coal or coke, burns, it continues to do so whilst oxygen is available, giving out heat in the process.

Substances burning in air combine with oxygen in definite proportions by mass. The substances do not normally combine with nitrogen, or any other of the constituent gases of the air. When burning commences, the atoms of the substance combine with the oxygen atoms in the oxygen molecules contained in the air. For example:

(1) Carbon (chemical symbol C, C is also used to represent a carbon atom) combines with two oxygen atoms (i.e. an oxygen molecule), the burning reaction being represented by the chemical equation

$$C + O_2 \rightarrow CO_2 \qquad (1)$$

Equation (1) is essentially a shorthand way of stating that one atom of carbon combines chemically with two atoms (or one molecule) of oxygen to form the compound, a gas in this case, of carbon dioxide. Incidentally, when carbon burns to form carbon dioxide (CO_2), it gives out 33×10^6 J of heat for each kilogram consumed. (2) Magnesium (Mg) combines with oxygen atoms on a one-to-one basis, so two Mg atoms combine with two oxygen atoms or one oxygen molecule, the chemical equation being

$$2Mg + O_2 \rightarrow 2MgO \qquad (2)$$

Note that we do not write the chemical equation as $Mg + O \rightarrow MgO$, since the smallest entity of oxygen which can have an independent existence is the oxygen molecule, i.e. O_2, not the oxygen atom, O. Equation (2) expresses in shorthand form the chemical reaction involved: two atoms of magnesium combine with one molecule of oxygen to form the compound magnesium oxide. Note that the chemical formula of a substance expresses the composition of the substance in terms of the atoms of the elements it contains. In the actual reaction, of course, millions and millions of atoms and molecules are involved. However, the reaction proceeds such that one magnesium atom always combines with one oxygen atom; and thus when magnesium oxide is produced in the chemical reaction, there is a definite rearrangement of the atoms of magnesium and oxygen.

1.5
The description of an oxide as a compound of an element and oxygen

When an element, such as carbon, copper, magnesium, lead, sulphur, or phosphorus, burns in oxygen, it combines chemically with the oxygen to form a compound. The resulting compound is known as an oxide.

In general, the compound of any element and oxygen is known as an oxide. Oxides may be in the form of a gas, liquid, or solid, although they mainly occur in the solid state. For example, when carbon burns in air, it forms the compound of carbon dioxide and, in some cases, some carbon monoxide (CO) as well; when hydrogen combines chemically with oxygen, it forms water (H_2O); the compound of magnesium and oxygen, magnesium oxide (MgO), is a solid.

1.6
The description of how oxygen and water are involved in rusting and the recognition that rusting is a chemical reaction

Let us first define the meaning of the term rusting. Rusting is the process in which iron changes chemically to a complex compound of iron oxide together with attached water molecules. Rust itself has the approximate formula $2Fe_2O_3 . 3H_2O$; Fe_2O_3* is an iron oxide known as ferric oxide. Rust has the chemical description of hydrated ferric oxide, 'hydrated' to qualify the presence of water molecules. Iron will not rust in perfectly dry air, nor will iron rust in water which is completely free from dissolved oxygen.

The fact that oxygen and water are involved and indeed required for rusting can be verified experimentally as follows:

(1) Place an iron or steel nail in a perfectly dry atmosphere of air, as shown for example in fig. 1.2(a). The concentrated sulphuric acid acts as a dehydrating agent (absorbs any water vapour) thus ensuring that no water is present in the air in the beaker. Observe that no rusting

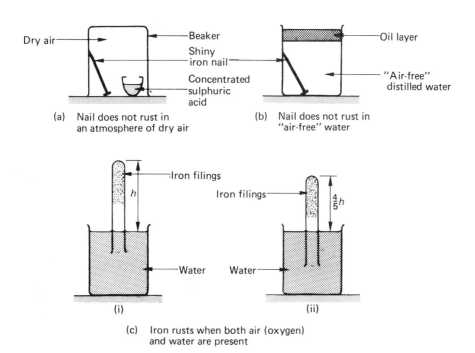

(a) Nail does not rust in an atmosphere of dry air

(b) Nail does not rust in "air-free" water

(c) Iron rusts when both air (oxygen) and water are present

Fig. 1.2

* Fe is the chemical symbol for iron; H_2O is the chemical formula for water; Fe_2O_3 for ferric oxide.

occurs even over periods of a week or more by the fact that the nail remains in its original shiny state. Rusting would be indicated by the nail going a reddish-brown colour.

(2) Place some distilled water in a beaker and boil to expel any air that may have been absorbed in the water. Immerse a nail in the water and cover water surface with a layer of oil to ensure no air comes into contact with the water as shown in fig. 1.2(b). This layer prevents any possibility of air dissolving in the water. Observe that no rusting occurs.

(3) Wet the sides of a test tube with water. Place some iron filings in the tube and shake so that they adhere to the tube sides. Place the tube, mouth downwards, in a beaker of water, as shown in fig. 1.2(c)(i). Leave for approximately one week. After this time it will be observed that the filings will have rusted, and if the water levels in the beaker are equalised (so that 'air' in the tube is at atmospheric pressure) it will also be observed that the 'air' in the tube is reduced by one-fifth. This indicates that one-fifth by volume of the original air in the tube has been used up in the rusting process. This one-fifth is oxygen.

From observations 1, 2 and 3 we can conclude that both oxygen and water are necessary for rusting to occur.

1.7
A discussion of the damage done by rusting and methods used to prevent rusting

Figure 1.3 shows an example of the damage done by the rusting of a steel plate with one surface exposed to air and water. The outer layers of rust are only loosely connected to the layers nearest the yet unrusted steel. The inner rust layers adhere firmly to the steel. As the corroding process of rusting continues the steel plate is weakened and will eventually fracture if under pressure, or rust through completely.

It is estimated that the annual costs of metal corrosion amount to between £20 and £50 per person in an industrialised country. In Britain approximately £1300 million (about 3.5 % of our gross national product) is lost annually by corrosion. A very high percentage of this is due to the rusting of iron and steel. It is also estimated that between 10 and 20 % of the production of iron and steel is used to replace iron lost by rusting.

Immense amounts of money and effort are therefore spent to combat and reduce rusting. The simplest means of preventing rusting is to ensure that air (oxygen) and water, the reactants necessary for rusting, are stopped from reaching the iron. Thus rusting can be prevented by greasing or painting exposed iron surfaces. Plating of steel surfaces with another metal is also extensively used. For example zinc plating, known as galvanising, produces a protective layer of zinc on steel; the zinc subsequently reacts with the oxygen and carbon dioxide in the air to form a strong adherent coating which is very resistant to corrosion.

Problems 1: Basic chemical processes and reactions

1. (a) Write down the two main constituents of air.
(b) Describe an experiment to demonstrate that a metal gains mass when heated in air. Explain why the metal gains mass.

2. (a) Describe the meaning of the term oxide and give one example each of an oxide which is in a gaseous, a liquid, and a solid state at normal temperatures.
(b) Describe the process of carbon burning in air. Include in your answer the chemical equation of the reaction.

3. What two constituents are necessary for the rusting of iron to occur? Describe an experiment to show that both constituents are necessary.
Comment on the effects of the damage caused by rusting.

4. Describe what is meant by a chemical reaction. Give two examples of a chemical

Surface exposed to air and water

Unrusted steel plate

Rusting steel plate

"Flakey" or loose rust

Adherent rust

Steel

Fig. 1.3

reaction, including in one the chemical equation of the reaction.

5. (a) Define what is meant by an oxide.
(b) With what element does a substance burning in air combine?

(c) Give examples of the damage done by rusting.

(d) Describe one method used to prevent rusting.

2   Elasticity of materials: Hooke's Law

The expected learning outcome of this chapter is that the student should be able to describe the elasticity of materials in terms of Hooke's Law.

Introduction: (a) tensile forces and stresses, (b) compressive forces and stresses, (c) shear forces and stresses, (d) elasticity

There are three main types of internal forces—tension, compression, and shear—which may be set up in a body or in the components of a system which is acted upon by external forces. The internal forces are known as forces of stress. Stress itself is defined in terms of the force per unit area produced in the material by the external force, i.e.

$$\text{stress} = \frac{\text{external applied force}}{\text{area over which force acts}}$$

The SI unit of force* is the newton, symbol N, and thus the units of stress are newtons per square metre, N/m^2.

(a) Tensile forces and stresses

Where the action of the external force tends to stretch or pull apart the component, the component is said to be under tension. The force producing the tension is known as a tensile force.

In fig. 2.1(a), a bar is clamped at end B and its other end C is subjected to a pulling force, i.e. a tensile force. The bar will extend and will be in a state of tension. Inside the bar itself there are internal forces which resist the applied force. These internal forces are cohesive forces between the grains and within the grains of the bar material and their cumulative effect per unit area is called stress; in this case, tensile stress since the bar is in a state of tension. The magnitude of the stress is determined by dividing the applied force by the cross-sectional area of component, i.e.

$$\text{tensile stress} = \frac{\text{tensile force}}{\text{area}} = \frac{F}{A}$$

In the definition of a tensile stress, the area A under consideration is at right angles to the line of action to the applied force, F.

Other examples of components undergoing tensile stress are shown in fig. 2.1(b) and (c), where the tensile stresses in the rope and cable components are identified by two opposing arrows.

(b) Compressive forces and stresses

In fig. 2.2(a), the bar is subjected to a pushing force which tends to compress the bar. This creates a compressive stress in the bar and the magnitude of the compressive stress is given by

$$\text{compressive stress} = \frac{\text{force creating compression}}{\text{area at right angles to force}}$$
$$= \frac{F}{A}$$

In fig. 2.2(b) and (c), the spring S and supports X and Y are in compression, whilst in (d) all components, except AB and BC, are undergoing compressive stress. The vertical load-bearing walls of a building and the piers of bridges are further examples of components undergoing compressive stress.

(c) Shear forces and stresses

In cases where the applied force is at right angles to the cross-sectional area of the component, the force produces either a tensile or compressive stress in the component. Another type of stress, known as shear stress, occurs when the plane of

---

* The newton is defined in Section 5. It is difficult to give a rigorous definition of force itself at this point. Force may be defined as the quantity which changes or attempts to change the condition of rest of a body or its state of motion. At present we may regard an applied force as a quantity measured in newtons which effects extension, compression, or shear in a material. Internal forces are set up in the material to counterbalance or work in opposition to an applied force.

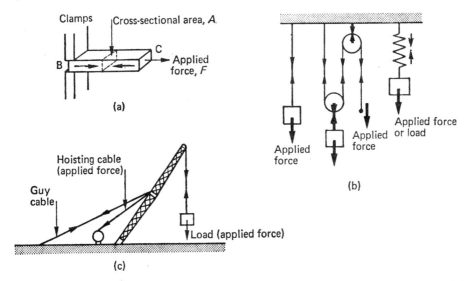

Fig. 2.1   Diagrams showing components under tension. Tensile stress in the individual components is indicated by → ←

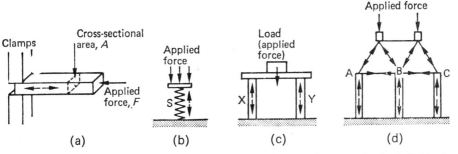

Fig. 2.2   Diagrams showing components under compression. Compressive stress in the individual component indicated by → ←

an area under stress is in the same direction as the applied force.

Examples of components undergoing shear stresses are shown in fig. 2.3. In (a) the base of the cube is clamped and a force is applied parallel to the top face of the cube. The cube undergoes distortion in which layers of the cube material tend to slide over adjacent layers. In (b) the rod is clamped at its base and shear stresses across the various cross-sectional areas of the rod are produced by a torsional or twisting force applied at the uppermost section of the rod. (c) shows an example of a rivet undergoing a shear stress. The rivet is to keep one plate from sliding over the other as well as to secure the plates together. A shear stress which actually causes eventual fracture is set up in a material when cut by hand shears or by means of a guillotine blade.

(d) *The definition of elasticity*

Provided that the distortion produced by an applied force is relatively small, most materials return to their original shape when the applied force is removed. This capacity of materials to return to their original shape when applied forces are removed is called elasticity.

2.1
The statement of Hooke's Law and the experimental determination of the relationship between force and extension for different materials

In order to give a quantitative measure of the distortion in a body or material when subjected to applied forces, we introduce the term strain.

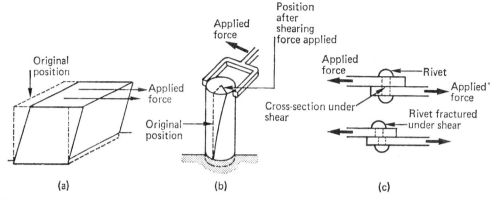

Fig. 2.3   Diagrams showing components under shear stress

Strain is a measure of the ratio of the distortion produced to an original dimension. The three strains corresponding to tensile, compressive, and shearing forces, are defined as:

$$\text{tensile strain} = \frac{\text{extension}}{\text{original length}} = \frac{x}{L},$$

see fig. 2.4(a)

$$\text{compressive strain} = \frac{\text{compression}}{\text{original length}} = \frac{x}{L},$$

see fig. 2.4(b)

$$\text{shear strain} = \frac{\text{shear displacement}}{\text{original length}} = \frac{x}{L},$$

see fig. 2.4(c)

Strain is the ratio of two lengths and therefore has no units.

Now having established the meaning of stress and strain, we can consider the relation between them for materials having the property of elasticity. The relation between the extension produced by a tensile force was investigated by

Robert Hooke, who found that as the applied force was gradually increased, the extension produced was at first proportional to the value of the applied force. Thus, if we were to plot a graph of applied force (also termed load) against extension, we would obtain a straight line which passes through the origin. A value of loading, however, can be found beyond which this would no longer be true. This point is called the limit of proportionality. These findings may be expressed in the following statement of Hooke's Law:

the strain produced in an elastic material is directly proportional to the applied stress, provided that the limit of proportionality has not been exceeded.

Hooke's Law is closely obeyed by many materials, provided that the strain is relatively small. For example, for most metals the limit of proportionality corresponds to a maximum strain of the order of 0.1%. There are also materials which have the property of elasticity in that they return to their original shape when the load is removed but do not have a straight-line stress versus strain graph.

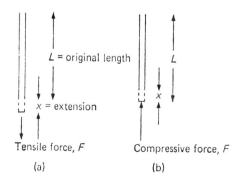

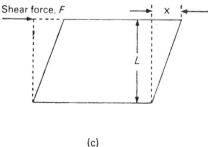

Fig. 2.4

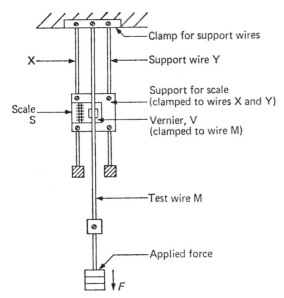

Fig. 2.5

length $L = 1\,m$ of mild steel wire of radius $r = 0.564\,mm = 5.64 \times 10^{-4}\,m$ was loaded:

Load F (N)	Extension x (mm)	Stress F/A (N/m²)	Strain x/L
0	0	0	0
40	0.189	40×10^6	0.189×10^{-3}
80	0.377	80×10^6	0.377×10^{-3}
120	0.566	120×10^6	0.566×10^{-3}
160	0.755	160×10^6	0.755×10^{-3}
200	1.04	200×10^6	1.04×10^{-3}
240	1.42	240×10^6	1.42×10^{-3}
260	1.80	260×10^6	1.80×10^{-3}
280	2.26	280×10^6	2.26×10^{-3}
300	2.80	300×10^6	2.80×10^{-3}

Figure 2.5 shows an experimental set-up which may be used to investigate the load versus extension relation of a metal wire. Wire M is the test specimen, whilst wires X and Y are used to carry scale S. The vernier V is clamped to M and is used to measure the extension when M is loaded. All three wires are clamped to a common support. Thus if the support were to sag when loads are applied to M, the scale would also drop by an equal amount and hence annul any false extension caused by the support sag.

The following results were obtained when a

The load values F were converted to stress by dividing each value by the cross-sectional area of the wire, i.e. $A = \pi r^2 = 3.1416 \times 5.64 \times 10^{-4} \times 5.64 \times 10^{-4} = 10^{-6}\,m^2$, and the extension values were converted to strain by dividing each value by $L = 1000\,mm$. Graphs of load versus extension and stress versus strain are plotted in fig. 2.6(a) and (b) respectively. Note that the latter deviates from a straight line for strains greater than 0.8×10^{-3} (stresses $> 170 \times 10^6$ N/m²) and so Hooke's Law is only obeyed for strains up to 0.08 %, and the stress corresponding to the limit of proportionality of M is 170×10^6 N/m².

In the above experiment we actually exceeded the elastic limit of material M since—although

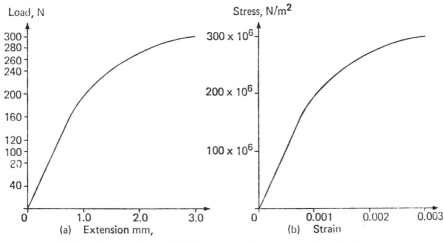

(a) Load v extension and (b) Stress v strain graphs for a 1 m length
of mild steel wire of diameter 1.128 mm

Fig. 2.6

no results are given—on unloading M it was observed that M was left with a permanent extension; the elastic limit of a material being defined as the maximum loading which, when removed, allows the material to recover to its original dimensions.

The ratio of stress to strain up to the limit of proportionality of a material which obeys Hooke's Law is a constant. This constant is known as the modulus of elasticity, and for tensile and compressive stress the constant is usually known as Young's modulus, denoted by E:

$$\text{Young's modulus } E = \frac{\text{stress}}{\text{strain}} = \frac{F/A}{x/L} = \frac{FL}{xA} \text{ N/m}^2$$

and since strain has no units, E has the units of stress, i.e. N/m^2. Some average values of E for some common materials are given below in Table 2.1. Values of tensile strength which is defined as the maximum tensile stress beyond which the material would eventually break are also included for interest.

2.2
The solution of simple problems involving Hooke's Law

Remember:

$$\text{stress} = \frac{\text{applied force or load}}{\text{cross-sectional area}} \text{ N/m}^2$$

$$\text{strain} = \frac{\text{extension or compression}}{\text{original length}}$$

$$\frac{\text{stress}}{\text{strain}} = \text{constant}$$

$\qquad = E$, Young's modulus for tension and compression, up to the limit of proportionality

Example 2.1
A tensile force of 50×10^3 N when applied to a brass rod of length 50 cm and cross-sectional area 9 cm² produces an extension of 0.29 mm. A compressive force of the same value produces a compression of 0.29 mm. Assuming that the limit of proportionality is not exceeded, calculate:

(a) the extension produced by a load of 70×10^3 N,

Table 2.1 Young's modulus and tensile strength of some common materials

Material	Young's modulus E (N/m²)	Tensile strength (N/m²)
Aluminium	68 to 72 × 10⁹	70 to 150 × 10⁶
Copper	96 to 132 × 10⁹	120 to 400 × 10⁶
Mild steel	190 to 220 × 10⁹	420 to 510 × 10⁶
Iron (cast)	90 to 130 × 10⁹	100 to 230 × 10⁶
Iron (wrought)	180 to 195 × 10⁹	280 to 450 × 10⁶
Concrete	~28 × 10⁹	~4 × 10⁶
Glass (crown)	~70 × 10⁹	30 to 90 × 10⁶
Oak (along grain)	~11 × 10⁹	60 to 110 × 10⁶
Nylon (at 20 °C)	0.8 to 3.1 × 10⁹	76 to 97 × 10⁶

(b) the compression produced by a compressive force of 20×10^3 N.

Determine also the value of Young's modulus for brass.

Solution
Stress $\propto$ strain, load $\propto$ extension (or compression) for a given component. In the above example we know that a load of 50×10^3 N produces an extension (or compression) of 0.29 mm, so we have

$$50 \times 10^3 = k \times 0.29 \quad \text{where} \quad k \text{ is a constant}$$

(a) Let extension produced by 70×10^3 N load be x mm, then

$$70 \times 10^3 = kx$$
but
$$50 \times 10^3 = k \times 0.29$$

and on dividing the two equations to cancel k we obtain

$$\frac{70}{50} = \frac{x}{0.29} \quad \text{so} \quad x = \frac{7}{5} \times 0.29 = 0.406 \text{ mm}$$

(b) Let compression produced by 20×10^3 N load be x', then

$$20 \times 10^3 = kx'$$
but
$$50 \times 10^3 = k \times 0.29$$
so
$$x' = \tfrac{2}{5} \times 0.29 = 0.116 \text{ mm}.$$

$$\text{Young's modulus } E = \frac{\text{stress}}{\text{strain}} = \frac{F/A}{x/L} = \frac{FL}{xA}$$

The cross-sectional area of the brass rod $A = 9 \text{ cm}^2 = 9 \times 10^{-4} \text{ m}^2$, and an applied force

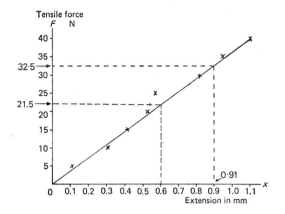

Fig. 2.7

$F = 50 \times 10^3$ N produces an extension of $x = 0.29$ mm in an original length $L = 50$ cm $= 500$ mm, so we have for brass,

$$E = \frac{(50 \times 10^3) \times (500)}{(0.29) \times (9 \times 10^{-4})} = \frac{25 \times 10^6}{9 \times 0.29 \times 10^{-4}}$$

$$= \frac{25}{9 \times 0.29} \times 10^{10} = 9.58 \times 10^{10} \text{ N/m}^2$$

Example 2.2
The following experimental results were obtained from a tensile loading test on a length of steel wire:

Tensile force, F (N)	0	5	10	15	20
Extension, x (mm)	0	0.12	0.3	0.42	0.54

Tensile force, F (N)	25	30	35	40
Extension, x (mm)	0.6	0.84	0.96	1.12

Plot a graph of tensile force against extension and draw the 'best-fit' straight line through these points. Determine from your graph:

(a) the tensile force which produces an extension in the wire of 0.6 mm,
(b) the extension produced by a tensile force of 32.5 N.

Note: The accuracy of the equipment used to measure both F and x is better than $\pm 5\%$.

Solution
The points are plotted in fig. 2.7. The 'best-fit' straight line is drawn in so that this line passes close to or through the points. It is the best 'compromise' allowing for the fact that experimentally measured values of F and x have a

$\pm 5\%$ tolerance. Clearly the point corresponding to $x = 0.6$ mm, $F = 25$ N is in error (and in a practical situation this point should be remeasured).

(a) To find the tensile force F which produces an extension of $x = 0.6$ mm draw a vertical line through $x = 0.6$ mm to cut the 'best-fit' straight-line graph. Then draw horizontal line through this point of intersection and read off F from tensile force axis. These lines are shown dashed in fig. 2.7 and give $F = 21.5$ N when $x = 0.6$ mm.
(b) Similarly, to find x when $F = 32.5$ N, draw a horizontal line through $F = 32.5$ N to cut the straight-line graph. Then draw a vertical line through the point of intersection and read off the value of x from the extension axis. These lines are shown as dashed lines in fig. 2.7 and give $x = 0.91$ mm when $F = 32.5$ N.

Problems 2:   Elasticity of materials: Hooke's Law

1.   Define elasticity and state Hooke's Law. Describe a simple experiment to investigate the relationship between the extension of (a) a strip of rubber of uniform cross-section and (b) a metal wire when subjected to gradually increasing loads. Sketch graphs of extension versus load for both cases, illustrating the results you might expect.
2.   A metal bar of 2 m length and cross-sectional area 10 mm^2 is extended 5 mm when subjected to a tensile force of 250 N. Calculate (a) the stress in the bar and (b) the extension if the tensile force is increased to 400 N.
3.   Given that rod A of a material extends 0.3 mm when subjected to a load of 40 N and rod B extends 0.4 mm when also loaded by 40 N, calculate:
(a) The extension of rod A when loaded by 60 N.
(b) The extension of rod B when loaded by 10 N.
(c) The load required to extend rod A by 0.5 mm.
    If rod A is joined end-to-end with rod B, calculate the extension of the combination when loaded by 80 N. It is assumed in all cases that the rods are not extended beyond their limit of proportionality.
4.   A steel cable of cross-sectional area 400 mm^2 carries a load of 20 kN. Calculate the tensile stress in the cable. If the maximum allowable working stress is 70 MN/m^2 (7×10^6 N/m^2), determine the minimum cross-sectional area of

cable that could be safely used to support the 20 kN load.

5.   Define stress and strain. A tensile force of 75×10^3 N produces an extension of 0.35 mm in a metal bar of unstressed length 0.4 m and a cross-sectional area 100 mm^2. Calculate, assum-ing that the bar is not stressed beyond its limit of proportionality:

(a) The extension produced by a force of 90×10^3 N.

(b) The tensile stress in the bar under a load which produces an extension of 0.45 mm.

3 Statics

The expected learning outcome of this chapter is that the student solves problems involving co-planar forces in static equilibrium situations.

3.1
The distinction between scalar and vector quantities and examples of such quantities

The definition of a scalar quantity

A scalar quantity is a quantity which possesses magnitude only.

Examples of scalar quantities are:

mass e.g. 10 kg; volume e.g. 10 m³; density e.g. 1000 kg/m³; time e.g. 30 s; temperature e.g. 0°C, 373 K, −39°C; speed e.g. 341.9 m/s; energy and work e.g. 2040 J; power e.g. 1000 W.

The definition of a vector quantity

A vector quantity is a quantity which possesses both magnitude and direction.

Examples of vector quantities are: velocity, acceleration, force, electric fields, magnetic fields.

Vector quantities may be represented on a diagram by means of a straight line of a given length and drawn in a given direction. The length of the line represents the magnitude of the quantity and the line direction, usually marked with an arrowhead, the direction in which the quantity acts. For example, a velocity v_1 of 2.0 m/s in the direction of a fixed line AN may be represented on a diagram by drawing a line of length |20 mm (scale 10 mm = 1 m/s) in the direction AN as shown in fig. 3.1(a); (b) represents a velocity v_2 of magnitude 3.0 m/s and in a direction at 30° to AN; and (c) represents a velocity v_3 of magnitude 1.10 m/s in a direction at 90° to AN. A convenient means of specifying vectors is to write down their magnitude followed by their angle to a given fixed direction. Thus the velocities in fig. 3.1 may be written as:

$$v_1 = 2.0 \, \text{m/s} \angle 0°, \quad v_2 = 3.0 \, \text{m/s} \angle 30°,$$
$$v_3 = 1.1 \, \text{m/s} \angle 90°.$$

3.2
Force is a vector quantity

Force is a vector quantity. A force must be defined by giving

(1) its magnitude in newtons,
(2) its line of action, i.e. the direction in which the force 'pulls' or 'pushes',
(3) its point of application.

For example in fig. 3.2, four forces F_1, F_2, F_3, and F_4, acting at point O, have been drawn. Point O is the point of application. A scale of 10 mm = 20 N is used and line OX is chosen as the reference direction, e.g. a force of 20 mm length in direction OX would represent the force

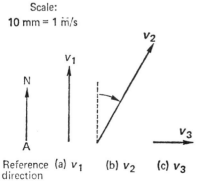

Fig. 3.1 Velocity vectors

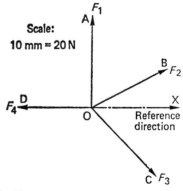

Fig. 3.2 Force vectors acting at point O

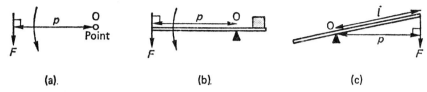

Fig. 3.3   Moment of force about a point = Force × perpendicular distance

40 N $\angle 0°$. Using a ruler and protractor determine the magnitude and direction of the four forces. The answers are:

$F_1 = 48$ N $\angle 90°$ (length OA = 24 mm, so magnitude of F_1 is $20 \times 24 \div 10 = 48$ N, F_1 makes $\angle$ AOX $= \angle 90°$, with reference direction OX), $F_2 = 46$ N $\angle 27°$, $F_3 = 51$ N $\angle 315°$, $F_4 = 40$ N $\angle 180°$.

Note that all the angles associated with specifying the direction of the forces are measured from OX in the counter-clockwise direction. By convention this direction is taken as positive. Angles measured in the clockwise direction are taken as negative angles. Thus F_3 may be quoted, as above, as 51 N $\angle 315°$, or alternatively as 51 N $\angle -45°$.

3.3
The definition of the moment of force about a point

The moment of force about a point is defined as the product of the magnitude of the force and the perpendicular distance of the line of action of the force to the point. Thus referring to fig. 3.3:

the moment of the force F about the point O
$$= F \times p \text{ newton-metres (Nm)}.$$

In (a) and (b) the moment is in a counter-clockwise direction, in (c) the moment is in a clockwise direction. If $F = 15$ N and $p = 3$ m, the magnitude of the moment, $F \times p = 15 \times 3 = 45$ Nm. Note that p must always be taken as the perpendicular distance from O to the line of action of the force; so in (c) we draw p *not* along the rod but as the line from O at right angles to the line of action of F.

3.4
The principle of moments

A body or a system consisting of a number of bodies and components is in equilibrium, that is

at rest, if two conditions are satisfied:

(1) The resultant of the forces acting on the body or system is zero.
(2) The sum of the clockwise moments about any point equals the sum of the counter- or anti-clockwise moments about the same point.

Condition 2 is the statement of the principle of moments. When 2 is satisfied there will be no turning or rotational motion. This principle is used to solve equilibrium problems when we are dealing with forces which do not all act at the same point. For example, we apply 2 for finding the equilibrium condition in loaded beam problems. When 1 is satisfied there is no translational motion, i.e. no motion in a straight line. The application of the equilibrium conditions is illustrated in the following example.

Example 3.1
Parallel forces of $F_1 = 10$ N, $F_2 = 48$ N, $F_3 = 38$ N, act on a rod in directions and positions as shown in fig. 3.4. Show that the rod is in equilibrium.

Solution
Forces F_1 and F_3 acting 'upwards' are balanced by F_2 acting downwards, i.e.

$$F_1 + F_3 = F_2, \quad 10 \text{ N} + 38 \text{ N} = 48 \text{ N}$$

However, we must also apply the principle of moments to check that there is no rotational motion. We can choose any point, so let us first try the centre-point O of the rod. If we consider the rod as pivoted at O, then the forces F_1 and F_2 will tend to rotate it in a clockwise direction, while F_3 would give an anti-clockwise rotation. If the rod is in equilibrium, then the clockwise

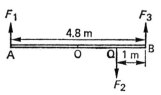

Fig. 3.4

moments must exactly balance the anti-clockwise one. Thus

clockwise moment about O
$$= F_1 \times AO + F_2 \times QO$$
$$= 10 \times 2.4 + 48 \times 1.4 = 91.2 \, \text{Nm}$$
anti-clockwise moment about O
$$= F_3 \times BO = 38 \times 2.4 = 91.2 \, \text{Nm}.$$

Hence total clockwise moment about O = anti-clockwise moment and thus the rod is in equilibrium. Note that if we took moments about Q, the moment due to F_2 would be zero and we would be left with:

clockwise moment $= F_1 \times AQ = 10 \times 3.8 \, \text{Nm}$
anti-clockwise moment
$$= F_3 \times BQ = 38 \times 1 \, \text{Nm}.$$

The two moments are equal in magnitude, exactly as we must expect. We need only apply the principle of moments for one point. If it holds for one point it will hold for any other. If it fails for one point it will fail for all other points.

3.5
The solution of simple problems using the principle of moments

Example 3.2
(a) Figure 3.5(a) shows a diagram of a simple lever system. Calculate the minimum value of the force F required to raise the 50 kg mass. Take $g = 9.81 \, \text{m/s}^2$.
(b) The beam shown in fig. 3.5(b) is pivoted at end P. The length of the beam is 6 m and its mass, which may be considered as concentrated at its

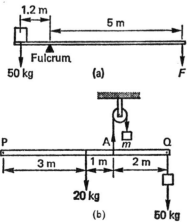

Fulcrum
50 kg   (a)   F

P   A m   Q
    3 m   1 m   2 m

20 kg

(b)   50 kg

Fig. 3.5

mid-point, is 20 kg. A mass of 50 kg is suspended at the far end Q of the beam. One end of a nylon-cord rope is attached to the beam at A, 2 m from Q. The cord rises vertically from the beam and is passed over a pulley, as shown, and a mass of m kg is secured to the far end. Neglecting any friction effects of the pulley system, calculate the value of m required to maintain equilibrium.

Solution
(a) If the force F is to raise the 50 kg mass, its moment about the fulcrum point must be at least equal to the counter-clockwise moment of the gravitational force of the 50 kg mass, i.e.

$$F \times 5 = 50g \times 1.2$$

so minimum value of

$$F = \frac{50g \times 1.2}{5} = 12g = 12 \times 9.81 = 117.7 \, \text{N}.$$

(b) For equilibrium the clockwise moments due to the gravitational forces of the 20 kg beam mass and the 50 kg mass about P must be balanced by the anti-clockwise moment due to mg acting at A about P. Thus

$$20g \times 3 + 50g \times 6 = mg \times 4$$

and, on cancelling g throughout,

$$60 + 300 = 4m, \quad \text{so} \quad m = \frac{360}{4} = 90 \, \text{kg}.$$

3.6
The definition of the centre of gravity

The centre of gravity of a body is that point through which the resultant of the earth's gravitational force of attraction upon the body can be considered as acting.

The total mass of a body is of course distributed throughout its volume. However, as far as the gravitation force on the body is concerned, we can consider the total mass to be concentrated at the centre of gravity point. This greatly simplifies force problems, e.g. in Example 3.2 we assumed that the mass of the beam could be considered to act at the beam centre, the centre point being the centre of gravity of a uniform beam. In fig. 3.6 the individual parts of the body shown experience downward forces due to gravitation attraction between the part and earth. The resultant gravitation force mg newtons acts through the centre of gravity of the

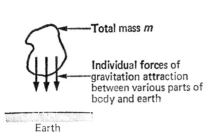

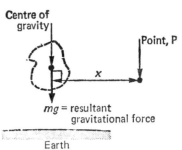

Fig. 3.6

body so, for example, the moment about a point P is mgx, where $x =$ perpendicular distance from P to line of action of force mg acting through centre of gravity of body.

The centre of gravity of a thin uniform rod is at the centre point of the rod, *see* fig. 3.7(a).

The centre of gravity of a rectangular lamina (a lamina is a thin plate or sheet of material) is at the intersection of the diagonals of the lamina, *see* fig. 3.7(b).

The centre of gravity of a circular lamina is at the centre of the circle, *see* fig. 3.7(c).

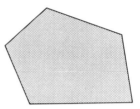

(a) Cardboard sheet

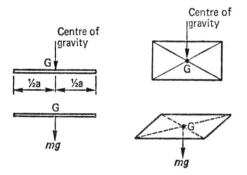

(a) Thin uniform rod   (b) Rectangular lamina

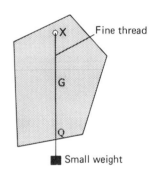

(b) Cardboard suspended at point X

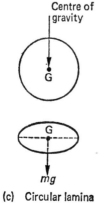

(c) Circular lamina

Fig. 3.7 Diagrams showing positions of centre of gravity

(c) Cardboard suspended at point Y

Fig. 3.8 Experimental determination of a sheet of material: the centre of gravity G lies at the point of intersection of the lines XQ and YR, where X and Y are the points of suspension

3.7
The experimental determination of the centre of gravity of an irregular sheet

When any body is suspended at a point such that it can freely move it will always come to rest with its centre of gravity lying vertically below the point of suspension. It is only in this way that the principle of moments can be satisfied.

This fact enables us to obtain experimentally the centre of gravity of sheets of material, however irregular their boundary, provided they are of constant thickness.

Consider, for example, the determination of the centre of gravity of a piece of cardboard such as that shown in fig. 3.8(a). Pierce the cardboard with a fine needle at any point fairly close to its boundary and mount the sheet vertically, as shown in fig. 3.8(b). Make sure the sheet is free to rotate in the vertical plane. In its equilibrium position its centre of gravity G will lie vertically below the point of suspension X, i.e. on line XGQ in fig. 3.8(b). This line can be marked on the cardboard as follows. From the needle at X suspend a fine length of thread with a small weight attached at its lower end and let it hang vertically very close to the cardboard. Mark two points on the line where the thread 'touches' the cardboard. Draw in the straight line XGQ.

Select a second point Y again close to the boundary and a good distance from X. Repeat the procedure to obtain the straight line YGR. The centre of gravity must lie on this line as well as line XGQ. Thus the centre of gravity G lies at the intersection of the two lines XGQ and YGR.

3.8
The description of stable, unstable, and neutral equilibrium

A body which remains at rest when acted upon by forces is said to be in equilibrium. The equilibrium of a body when acted upon by two or more forces may be stable, unstable, or neutral.

A body is said to be in stable equilibrium if, on receiving a small displacement (caused by a small 'push' or 'pull'), it tends to return to its original equilibrium position. For example, the bodies shown in fig. 3.9(a) are in a state of stable equilibrium. If subjected to a small displacement they will oscillate or 'rock' about their original position of equilibrium and return to this position when the oscillations have died down.

A body is said to be in unstable equilibrium if, on receiving a small displacement, it tends to increase the displacement and go farther away from its original position of rest. For example, the bodies shown in fig. 3.9(b) would tend to 'fall down' if given a small push, and are therefore in a state of unstable equilibrium.

A body is said to be in neutral equilibrium when it tends to rest in any position it is placed. The sphere, the cylinder, and the cone lying on its curved surface, *see* fig. 3.8(c), are examples of bodies in neutral equilibrium. If any one received a small push it would roll into a new position of rest.

Note that a cone is an example of a body that can be in the three different states of equilibrium, depending on how it is placed, i.e. stable on its base, unstable on its tip, and neutral on its side.

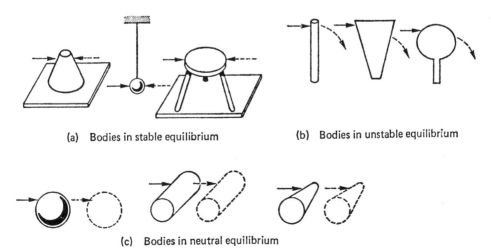

(a) Bodies in stable equilibrium

(b) Bodies in unstable equilibrium

(c) Bodies in neutral equilibrium

Fig. 3.9

Problems 3:   Statics

1.   Distinguish between a scalar and a vector quantity. Give three examples of each.

2.   Define the moment of a force about a point and state the principle of moments.

3.   Define the centre of gravity of a body and explain how this could be obtained experimentally for a sheet with an irregular periphery but of constant thickness.

4.   Distinguish between stable, unstable and neutral equilibrium. Draw diagrams showing a body in each of the three states.

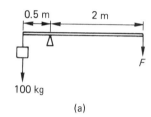

(a)

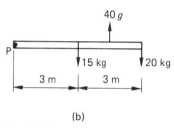

(b)

Fig. 3.10   for Problem 5

5.   (a) Calculate the magnitude of the force F in the simple lever system of fig. 3.10(a) which will just raise the 100 kg load. Take $g = 9.81 \text{ m/s}^2$, and neglect the mass of the levering rod.

(b) The rod in fig. 3.10(b) is pivoted at P. Its mass of 15 kg may be considered to act at its centre 3 m from P. If a mass of 20 kg is suspended at the far end of the rod (6 m from P), calculate the position of the point of application of the vertical $40g$ N force.

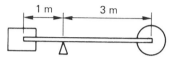

Fig. 3.11   for Problem 6

6.   Define the centre of gravity of a body and show with aid of a diagram the position of the centre of gravity of (a) a thin rod, (b) a rectangular lamina, (c) a circular lamina.

   Figure 3.11 shows a diagram of a thin uniform rod resting on a fulcrum with a circular lamina of mass 1.2 kg attached through its centre at one end and a rectangular lamina of mass 5.2 kg at the other end, also attached through its centre. If the rod is in equilibrium, calculate its mass.

Fig. 3.12   for Problem 7

7.   State the principle of moments. The rod in fig. 3.12 is loaded with masses of 8 kg and 5 kg. Calculate the moment of each mass about the pivot O. State in which direction the system will move. Neglect the mass of the rod. Determine the value of the mass which should be added to one end for equilibrium to be achieved. Take $g = 9.8 \text{ m/s}^2$.

4 Pressure in fluids

The expected learning outcome of this chapter is that the student knows the principles of pressure in fluids.

4.1
The definition of pressure

Pressure is defined as the normal force per unit area. By 'normal' we mean the force acting at right angles to unit area of the surface we are considering. This general definition describes quantitatively the force per unit area exerted on a surface by solids and fluids (fluids being the general term for both liquids and gases).

The solid block resting on the horizontal surface shown in fig. 4.1(a) exerts a vertical force on the surface. The force is normal, i.e. at 90°, to the surface. If the mass of the block is m kilograms, the gravitational force exerted by the block is mg newtons. If the area of the block resting on the surface is A square metres, then the pressure P,

$$P = \frac{\text{normal force}}{\text{area over which force acts}} = \frac{F}{A}$$

$$= \frac{mg}{A} \text{ newtons per square metre.}$$

A fluid in a container exerts forces on the container walls which are normal to these walls. For example the liquid in the tank shown in fig. 4.1(b) exerts a vertical force on the tank base and forces at right angles to the 4 vertical sides of the tank. If the total mass of liquid is m, the total

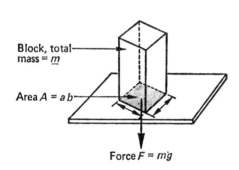

Block, total mass = m

Area $A = ab$

Force $F = mg$

(a) Total force over area A is F
Pressure $P = \frac{F}{A}$ = force per unit area, exerted over area A

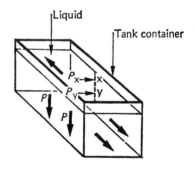

Liquid

Tank container

P_x ➤ x
P_y ➤ y
P

(b) Liquid in a tank: forces are exerted at 90° to tank walls. Pressure increases with depth, i.e. $P > P_y > P_x$

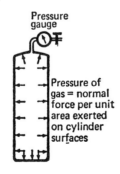

Pressure gauge

Pressure of gas = normal force per unit area exerted on cylinder surfaces

(c) Gas in a gas cylinder

Fig. 4.1

force exerted on the base is $F = mg$ and the pressure over the base area, of area A, is

$$P = \frac{\text{normal force}}{\text{base area}} = \frac{F}{A} = \frac{mg}{A} \ \text{N/m}^2.$$

The forces on the vertical sides of the tank increase with depth and thus the pressure on these sides also increases with depth reaching a maximum at the tank bottom. The pressure P_X at point X on a vertical side equals the force per unit area at point X. P_X will be less than the pressure P_Y at the deeper point Y, shown in the diagram.

The gas contained in the cylinder shown in fig. 4.1(c) exerts forces normal to the cylinder walls. The force per unit area exerted on the walls is the gas pressure.

Note that pressure is 'transmitted' throughout a fluid and its effects are experienced at the surfaces of the container or at the surfaces of a body immersed in a fluid.

The units of pressure are N/m^2 or pascals

Since pressure is force per unit area the units of pressure are newtons per square metre, abbreviation N/m^2.

The unit of pressure known as the pascal, abbreviation Pa, is also used: 1 Pa = 1 N/m^2.

4.2
The calculation of pressure, given force and area

Remember,

$$\text{pressure } P = \frac{\text{total force, } F \text{ newtons}}{\text{area, } A \text{ square metres}}$$

$$= \frac{F}{A} \ \text{N/m}^2 \text{ or Pa.}$$

Thus if a force of 1000 N acts over an area of 20 m^2,

$$\text{the pressure, } P = \frac{F}{A} = \frac{1000}{20} = 50 \ \text{N/m}^2.$$

If $F = 257$ N and $A = 50$ mm^2 = 50×10^{-6} m^2 (1 mm^2 = 10^{-6} m^2),

$$P = \frac{F}{A} = \frac{257}{50 \times 10^{-6}} = \frac{257}{50} \times 10^6$$

$$= 5.14 \times 10^6 \ \text{N/m}^2.$$

Let us also calculate the pressure of the block and the liquid shown in fig. 4.1(a) and (b), over their base areas:

(a) If we are given block mass $m = 200$ kg, and base dimensions $a = 0.25$ m, $b = 0.2$ m, we have (taking $g = 9.81$ m/s^2):

total force

$$F = mg = 200 \times 9.81 = 1962 \ \text{N,}$$

base area

$$A = ab = 0.25 \times 0.2 = 0.05 \ \text{m}^2,$$

so pressure

$$P = \frac{F}{A} = \frac{1962}{0.05} = 39\,240 \ \text{N/m}^2.$$

(b) If we are given area of tank base $A = 1.5$ m^2, volume of liquid in tank $v = 0.75$ m^3, and the density of the liquid $\rho = 870$ kg/m^3, we have

mass of liquid in tank = volume × density

i.e. $m = v \times \rho = 0.75 \times 870 = 625.5$ kg, so the total gravitational force exerted on base,

$$F = mg = 652.5 \times 9.81 = 6401 \ \text{N,}$$

and

$$\text{pressure, } P = \frac{F}{A} = \frac{6401}{1.5} = 4267 \ \text{N/m}^2.$$

4.3
The factors which determine the pressure at any point in a fluid: fluid density, depth, g, surface pressure

The factors that determine the pressure at any point in a fluid are:

(a) *Fluid density:* the greater the density of the fluid the greater the pressure.

For example, the pressure at the same positions below the surface of the liquid in the three vessels shown in fig. 4.2(a) and containing respectively water (density, $\rho_w = 10^3$ kg/m^3), turpentine (density, $\rho_t = 0.87 \times 10^3$ kg/m^3), and mercury (density, $\rho_m = 13.6 \times 10^3$ kg/m^3) are in the ratio $\rho_m : \rho_t : \rho_m = 1 : 0.87 : 13.6$.

(b) *Fluid depth:* the pressure increases with fluid depth.

For example, the pressure at points Q_2 in fig. 4.2 will be double the pressure at points Q_1, since pressure is directly proportional to the depth below the fluid surface.

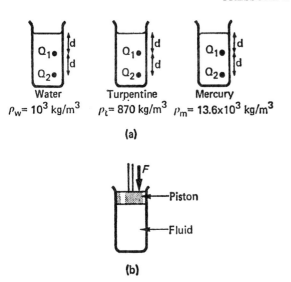

Water
$\rho_w = 10^3 \text{ kg/m}^3$

Turpentine
$\rho_t = 870 \text{ kg/m}^3$

Mercury
$\rho_m = 13.6 \times 10^3 \text{ kg/m}^3$

(a)

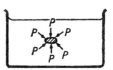

Fig. 4.3

(b)

Fig. 4.2

(c) *The acceleration due to gravity, g:* g enters pressure calculations to convert mass into a gravitational force, as we saw in the calculations of the previous sub-section.

(d) *Surface pressure:* a pressure applied at the surface of a fluid is transmitted throughout the fluid and will therefore increase the pressure at any point in a fluid by an amount equal to the surface pressure applied.

In fig. 4.2(b) a surface pressure is applied to the fluid surface by means of a tightly fitting piston. If the force applied to the piston is F and the area of the piston in contact with the fluid is A, then the piston exerts a surface pressure $P = F/A$ which is transmitted in all directions to all points in the fluid. In fact any open vessel, such as those in fig. 4.2(a), experiences atmospheric pressure as a surface pressure, so the total value of pressure at a depth below the surface of a fluid is the pressure due to the liquid plus atmospheric pressure (*see also* Section 4.7).

4.4
The pressure at any level in a liquid is equal in all directions

So far we have defined pressure as force per unit area and stated that pressure is transmitted in all directions throughout a fluid. If we were to place a tiny body in a liquid, as shown in fig. 4.3, and to neglect any variation of pressure with depth in the layer of liquid in which the body is situated, the body would experience equal pressure in all

directions. The pressure vectors act normally to the body's surface and are all equal in magnitude.

We can state generally that the pressure at any level in a liquid is equal in all directions.

The pressure at any point in a fluid is also independent of the shape of the vessel containing the fluid. Thus in fig. 4.4, where we have three

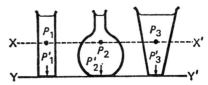

Fig. 4.4   Pressure is independent of shape of vessel

vessels of different shapes but filled with the same liquid to identical heights, we can state that the pressures at any point at the same depth below the liquid surface in each vessel are equal, e.g.

at level XX': $P_1 = P_2 = P_3$
at level YY': $P_1' = P_2' = P_3'$.

4.5
The pressure acts in a direction normal to its containing surface

The pressures exerted by a fluid on the surfaces of its containing vessel are always normal to these surfaces.

Examples showing that the directions at which pressure acts are at 90° (that is normal) to the containing surfaces are given in fig. 4.5.

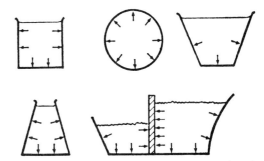

Fig. 4.5   Pressure acts in a direction normal to its containing surface

4.6

The pressure due to a column of liquid depends upon the density of the liquid and the height of the column

The pressure at any point in a liquid is directly proportional to the depth d of the point below the surface of the liquid and to the density ρ of the liquid. The formula for pressure, which we shall derive for the case of a column of liquid, is

$$P = d\rho g \text{ newtons per square metre}$$

where d = vertical depth below surface in metres; ρ = density of liquid in kilograms per cubic metre, kg/m^3; $g = 9.81\,m/s^2$, the acceleration due to gravity.

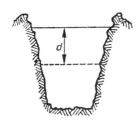

(a)   Pressure at depth d below surface is P = $d\rho g$

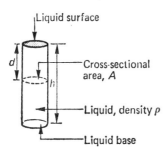

(b)   Column of liquid, pressure due to height h is P = $h\rho g$

Fig. 4.6

In the specific case of a uniform column of liquid of height h, density ρ, and cross-sectional area A (*see* fig. 4.6(b)) we can calculate the pressure at its base as follows:

total mass of liquid in column,

$$m = \text{volume of column} \times \text{density of liquid,}$$

so

$$m = (A \times h) \times \rho = Ah\rho \text{ kilograms,}$$

and total force acting on base,

$$F = mg = Ah\rho g \text{ newtons.}$$

Hence the pressure on base,

$$P = \frac{F}{A} = h\rho g \text{ newtons per square metre.}$$

The pressure at any level, depth d from the top of the column, is the force per unit area acting over the cross-sectional area A at depth d:

total force acting over A at d,

$$F_d = (\text{mass of column of height } d) \times g$$
$$= Ad\rho g,$$

so pressure at level d,

$$P_d = \frac{F_d}{A} = d\rho g.$$

Although the relations $P = h\rho g$ and $P_d = d\rho g$ were derived for a uniform column of constant cross-sectional area, they are quite general since pressure does not depend on the shape of the container. Pressure depends on the height or depth and the density of the liquid.

The solution of simple problems using $P = h\rho g$ or $d\rho g$

Example 4.1
(a) Atmospheric pressure is often quoted as 760 mm Hg, i.e. equivalent to the pressure exerted by a column of mercury of height 760 mm. Given that the density of mercury, $\rho = 13.59 \times |10^3\,kg/m^3$ and $g = 9.81\,m/s^2$, calculate the atmospheric pressure in pascals.
(b) Calculate the pressure exerted on a submarine which has dived to a depth of 200 m below sea-level. Take the density of sea-water, $\rho = 1025\,kg/m^3$, and $g = 9.81\,m/s^2$.
(c) A specially designed deep-sea diving vessel is constructed to withstand pressures of up to $10^8\,N/m^2$. Estimate the maximum depth to which the vessel could safely descend. Take mean density of sea-water as $1050\,kg/m^3$ and $g = 9.81\,m/s^2$.

Solution
(a) Atmospheric pressure,

$$P = h\rho g \text{ N/m}^2 \text{ or Pa,}$$

when $h = 760\,mm = 0.76\,m$ and $\rho = 13.59 \times 10^3\,kg/m^3$,

$$P = 0.76 \times 13.59 \times 10^3 \times 9.81$$
$$= 101.3 \times 10^3 \text{ Pa}$$

(b) Pressure at depth $d = 200$ m is

$$P = d\rho g = 200 \times 1025 \times 9.81$$
$$= 2.011 \times 10^6 \text{ N/m}^2$$

(c) Let d = maximum depth, then on equating maximum permissible pressure to the pressure at depth d, we have

$$10^8 = d\rho g = d \times 1050 \times 9.81 = 10\,300 d$$

so

$$d = \frac{10^8}{10\,300} = 10^4 \text{ m, approximately.}$$

Note that this result predicts that the vessel could operate farther below the surface of the sea than Mount Everest (height 8800 m) is above it. Such a vessel has to withstand enormous pressures and one indeed has been constructed with a steel casing of 125 mm thickness.

4.7
The distinction between absolute and gauge pressures

In this section we consider atmospheric pressure in a little more detail and define the terms absolute pressure and gauge pressure.

(a) Atmospheric pressure

The earth, as we well know, is surrounded by an atmosphere of air consisting mainly of the gases of nitrogen and oxygen, and we can therefore consider the earth's surface as being at the bottom of a deep ocean of air. The layers of air experience downward forces due to gravitational attraction of the earth and hence develop 'atmospheric' pressure exactly as pressure is developed by successive layers in a liquid. We can thus state that the atmospheric pressure at any given height above the earth's surface is due to the column of air above us. We should thus expect atmospheric pressure to be a maximum at sea-level and reduce with height above sea-level. These facts are indeed borne out by experience (see Table 4.1 below).

The average pressure the atmosphere exerts at sea-level, known as standard or normal atmospheric pressure, is 1.013×10^5 N/m². We have to consider an average value because atmospheric pressure varies with weather conditions. A

Table 4.1   Atmospheric pressure at fixed heights above sea-level

Height above sea-level (metres)	Atmospheric pressure (N/m²)	(mm of mercury)
0	1.013×10^5	760
1000	0.90×10^5	674
2000	0.80×10^5	597
3000	0.70×10^5	525
4000	0.61×10^5	462
5000	0.54×10^5	407

'dropping' pressure normally indicates poor weather; a rapidly dropping pressure, storms and gales. A rising pressure indicates a change to better weather conditions. An (atmospheric) pressure of 1.013×10^2 N/m² will support a column of mercury 760 mm in height, as shown in fig. 4.7. This provides a very convenient and accurate means of measuring atmospheric pressure in the mercury barometer. Hence atmospheric pressure is very often quoted as an equivalent height of mercury.

Atmospheric pressure, as is the case with all pressures in fluids, is transmitted equally in all directions and the fact that we do not 'feel' its pressure is due to this and the fact, fortunately for us, that an equal pressure acts on the inside of our bodies to balance it.

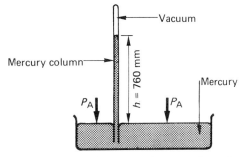

Fig. 4.7   Atmospheric pressure P_A supporting column of mercury. (Principle used in mercury barometer to measure P_A)

(b) Absolute and gauge pressures

In practice we may be interested in two kinds of pressure readings. Very frequently our main interest is the difference between the total pressure (including the effect of atmospheric pressure) and the pressure of the atmosphere on its own, i.e. the amount the total pressure either exceeds or is below atmospheric pressure. Most commercial–industrial pressure gauges register this, and hence the name gauge pressure is used to denote the difference between total pressure (absolute pressure) and atmospheric pressure:

gauge pressure = total or absolute pressure

− atmospheric pressure

The actual total pressure of a confined fluid including the effect of the 'outside' atmospheric pressure is known as the absolute pressure:

absolute pressure = gauge pressure

+ atmospheric pressure

4.8
The measurement of gas pressure using (a) a U-tube manometer, and (b) a pressure gauge

(a) The measurement of gas pressure using a U-tube manometer

Figure 4.8 shows a U-tube manometer being used to measure the pressure of a gas in container C. The U-tube contains a liquid. Mercury is most commonly used. In each diagram one end of the U-tube is open to the atmosphere and hence the surface pressure acting upon the liquid column in this side of the tube is the atmospheric pressure. We shall denote atmosphere pressure by P_A. The other side of the tube is connected to the container.

In (a) the levels of the liquid in the manometer are the same. Thus the absolute pressure of the gas P in the container exerted on the left-hand column must exactly balance the atmospheric pressure P_A exerted on the right-hand column of liquid, thus

$$P = P_A \, \text{N/m}^2$$

Note also the effect of inclining one limb of the U-tube: the levels of the manometer liquid are still the same. The gauge pressure is always determined by the vertical height difference between the two levels of liquid in the limbs of the U-tube.

In fig. 4.8(b) the pressure of the gas P_b must exceed atmospheric pressure, thus forcing the liquid in the U-tube round until the excess pressure caused by the height h of liquid plus atmospheric pressure equals P_b. Hence the absolute pressure of the gas in the container is

$$P_b = h\rho g + P_A \, \text{N/m}^2$$

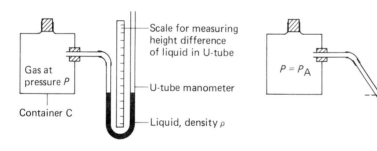

(a) $P = P_A$ (gas pressure equals atmospheric pressure)

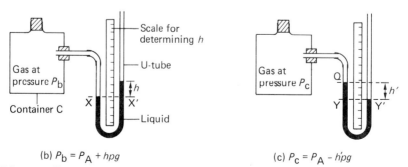

(b) $P_b = P_A + h\rho g$ (c) $P_c = P_A - h'\rho g$

Fig. 4.8   Diagrams illustrating the use of a U-tube manometer for measuring gas pressure

where h = vertical height difference of liquid level in U-tube;

ρ = density of manometer liquid, kg/m^3;

g = 9.81 m/s^2.

Alternatively we could use the following argument: the pressure at X′ in the liquid is $h\rho g$ above atmospheric pressure and since pressures at the same level of a liquid are equal, the pressure at X′ equals the pressure at X, which in turn equals the gas pressure P_b in container C.

In fig. 4.8(c) the gas pressure P_c in the container is below atmospheric pressure. The greater atmospheric pressure forces the liquid round in the U-tube until the excess pressure produced by the height h' of liquid causes the pressure on the two sides to equate. Hence the gas pressure in the container is

$$P_c = P_A - h'\rho g \, \text{N/m}^2$$

Alternatively we may argue: pressure at Y′ = pressure at Y = atmospheric pressure, but pressure at Y is $h'\rho g$ greater than at point Q due to the column of liquid; hence pressure at Q is $P_A - h'\rho g$, which is the pressure of the gas in the container.

The open-end U-tube described above can be used to measure pressures up to about twice atmospheric pressure. When $P = 2P_A$, $h = 760\,\text{mm}$ assuming mercury is used in the manometer. Thus higher pressures would require a length of U-tube in excess of 760 mm which is not very practical. The U-tube can be adapted to measure higher pressures by closing one end and calibrating the length of trapped air in this end in terms of pressure when the other end is attached to known pressures.

(b) Measurement of gas pressure using a pressure gauge

One of the most commonly used gauges employed to measure fluid pressure in industry is the Bourdon pressure gauge, an example of which is sketched in fig. 4.9. The pressure-sensing element consists of a flattened hollow metal tube, normally made from spring bronze or steel, which is bent into a circular shape. When the gauge is connected to measure gas or liquid pressure, the fluid enters the tube and the pressure it exerts tends to straighten the tube. The tube is fixed at the inlet side and the movement at the far end caused by the fluid

N/m^2 × 10^5

X

To measurement point

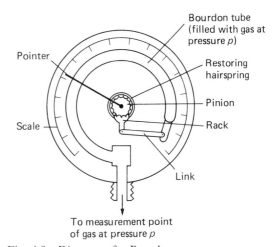

Pointer

Scale

Bourdon tube (filled with gas at pressure p)

Restoring hairspring

Pinion

Rack

Link

To measurement point of gas at pressure p

Fig. 4.9 Diagram of a Bourdon pressure gauge

pressure is communicated via a link to a rack and pinion mechanism which actuates a pointer. The pointer moves over a scale, calibrated to register pressure directly. For increased movement and therefore increased sensitivity the tube may be bent through several turns in the form of a spiral or helix.

Figure 4.10 shows sketches of two other gauges, which work on a similar principle. In the diaphragm gauge of (a) the fluid pressure acts on a diaphragm which is thus caused to press against the spring shown. The displacement of the diaphragm, which depends on the fluid pressure and the balancing force of the spring, is communicated via a rod to actuate a pointer which moves over a scale calibrated to read pressure directly. The diaphragm may be made of metal, for cases where high strength and resistance to corrosion are wanted, or of rubber if

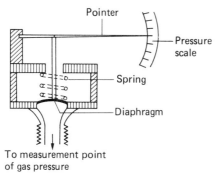

(a) A diaphragm pressure gauge

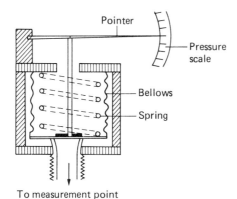

(b) A bellows pressure gauge

Fig. 4.10   Diagrams showing examples of two other
gauges used for measuring pressure

high sensitivity is required. The bellows-type gauge shown in (b) has the advantage of a greater range of deflection over the diaphragm gauge. The pressure exerted by the fluid at the base of the bellows is balanced by a spring and the displacement produced is transmitted via a rod to actuate a pointer.

There are many other types of pressure gauges available for a very wide variety of applications. At very low pressures encountered in high

vacuum work the McLeod gauge is frequently used. This gauge works on the principle of trapping a small amount of the gas whose pressure is required and then compressing this sample by a known amount and measuring the compressed sample pressure using a mercury U-tube manometer. The ratio of the actual to compressed sample gas pressure (i.e. the pressure measured by the manometer) is equal to the ratio of the compressed sample volume to the sample volume before compression. Using this technique pressures of the order of a millionth or lower of atmospheric pressure may be measured.

Problems 4:   Pressure in fluids

1.   Define pressure and state its units.
    A rectangular tank has a base area of 2 m × 3 m and is filled to a height of 2.5 m with a liquid of density 900 kg/m^3. Calculate (a) the force and the pressure exerted on the base, (b) the pressure on a tank side at a height 1.2 m above the base. Take $g = 9.81$ m/s^2.
2.   State the factors which determine the pressure at any point in a liquid.
    State whether the following statements are either TRUE or FALSE, giving a reason for your answer:
(a) Refer to fig. 4.11(a): the pressure on the base of vessel A is greater than the pressure on the base of vessel B.
(b) Refer to fig. 4.11(b): the directions of the resultant fluid pressure on the container walls are all correctly marked.
(c) Refer to fig. 4.11(c): the pressure at a depth 100 mm below the surface is greater in vessel C containing mercury (relative density 13.6) than 1 m below the surface in vessel D containing water.
(d) The pressure at any given point in a vertical

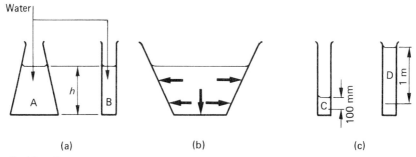

(a)   (b)   (c)

Fig. 4.11   for Problem 2

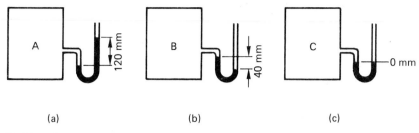

Fig. 4.12   for Problem 4

column of liquid is greater in the vertical direction than in a horizontal direction.

(e) Atmospheric pressure decreases with height above sea-level.

3.   Show that the pressure P at a depth d below the surface of a liquid of density ρ is given by $P = d\rho g + P_s$, where P_s = surface pressure on liquid, g = acceleration due to gravity.

   Calculate the pressure on the exterior frame of a submarine at a depth of 150 m below sea-level.

The density of sea-water is $1025\,kg/m^3$, atmospheric pressure is 10^5 Pa, and $g = 9.81\,m/s^2$.

4.   Describe how gas pressure may be measured using (a) a U-tube manometer, (b) a pressure gauge.

   Calculate the pressure of the gas in the containers A, B, and C shown in fig. 4.12. The liquid used in the U-tube is mercury, density $13.6 \times 10^3\,kg/m^3$. Take $g = 9.81\,m/s^2$ and atmospheric pressure as $101.3 \times 10^3\,N/m^2$.

B   Motion and energy

5   Dynamics: velocity, acceleration, distance–time problems

The expected learning outcome of this chapter is that the student solves problems involving distance, time, velocity and acceleration.

5.1
The definition of speed

Speed is the rate of change of distance with time. If the speed u is constant, i.e. does not change its value with time, and the distance travelled in a time of t seconds is s metres, then

$$\text{speed} = \frac{\text{distance travelled}}{\text{time taken}}, \quad u = \frac{s}{t} \text{ metres per second (m/s)}$$

$$\text{distance} = \text{speed} \times \text{time}, \quad s = u \times t \text{ metres}$$

$$\text{time} = \frac{\text{distance}}{\text{speed}}, \quad t = \frac{s}{u} \text{ seconds}.$$

If the speed changes with time, and therefore with distance travelled, then we define the average value of speed as

$$\text{average speed} = \frac{\text{total distance travelled}}{\text{total time taken}}.$$

The SI unit of speed is the metre per second, m/s, although kilometres per hour, km/h, may also be used.

5.2
The calculation of speed from given time and distance data

Example 5.1
(a) A particle travelling at a uniform speed in a straight line passes two points in this line at times of 2.4 s and 9.7 s. If the two points are 12.3 m apart, calculate the speed of the particle.
(b) Calculate the average speed of an athlete who runs (i) 100 m in 10.3 s, (ii) 5000 m in 13 minutes and 25 seconds.

(c) Calculate the average speed of a car taking 4 hours 30 minutes for a journey of length 280 km.

Solution

(a) $\text{Speed} = \dfrac{\text{distance travelled}}{\text{time taken}}$

$$= \frac{12.3}{9.7 - 2.4} = \frac{12.3}{7.3} = 1.68 \text{ m/s}.$$

(b) (i) $\text{Average speed} = \dfrac{100}{10.3} = 9.71 \text{ m/s}.$

(ii) First we must convert 13 minutes 25 seconds to seconds, i.e. $t = (13 \times 60 + 25) = 805$, so

$$\text{average speed} = \frac{5000}{805} = 6.21 \text{ m/s}.$$

(c) 4 hours 30 minutes $= 4.5\,\text{h}$ or $4.5 \times 60 \times 60 = 16\,200\,\text{s}$, so

$$\text{average speed} = \frac{280}{4.5} = 62.2 \text{ km/h}$$

or

$$\frac{280 \times 1000}{16\,200} = 17.3 \text{ m/s}.$$

5.3
Distance–time data and the plotting of distance–time graphs

The distance–time data of Table 5.1 were obtained

Table 5.1   Distance–time data for sports car test

	Time, t	Distance, s
A	0	0
B	30 min	40 km
C	1 h 0 min	100 km
D	1 h 30 min	180 km
E	2 h 30 min	250 km
F	3 h 30 min	300 km

for a sports car under test, the car being driven at constant speed during each time interval. The time taken for the car to accelerate or decelerate, to change its speed at the beginning of a new time interval, is neglected. The distance–time data of Table 5.2 were obtained by monitoring the motion of a certain point on a vibrating spring over a quarter cycle of its vibration.

Table 5.2   Distance–time (s–t) data for a point on a vibrating spring

	t (s)	s (mm)		t (s)	s (mm)
a	0	0	g	0.3	9.7
b	0.05	1.9	h	0.35	10.7
c	0.1	3.7	i	0.40	11.4
d	0.15	5.4	j	0.45	11.9
e	0.2	7.1	k	0.5	12.0
f	0.25	8.5			

Distance–time graphs for these data are plotted respectively in fig. 5.1(a) and (b). In both cases, time is plotted along the horizontal axis and the distance travelled along the vertical axis.

In (a) we take, for the time scale: 1 unit = 0.5 h (30 minutes); and for the distance scale, 1 unit = 50 km. The first point A: $t = 0$, $s = 0$, corresponds to the origin; the second point B: $t = 0.5$, $s = 40$ km, may be plotted by moving one unit along the time axis and then 0.8 units vertically corresponding to a distance of 40 km; the other 4 points C, D, E, F are plotted as shown. Since the car travels with constant speed in each of the given time intervals, the distance in these intervals increases uniformly with time. We can therefore join up the points A to B, B to C, C to D . . . etc. by straight lines.

In (b) we take for the time scale: 1 unit = 0.1 s; and for the distance scale, 1 unit = 2 mm. The points a, b, c, d, e, f, g, h, i, j, k corresponding to the data table are plotted as shown. This time we are not told that the speed is constant between time intervals, so we must therefore draw a smooth curve through the points.

5.4
The calculation of the gradient of distance–time graphs and the interpretation of the gradient as equal to speed

The gradient or slope at any point on a distance–time graph (with distance plotted on the vertical

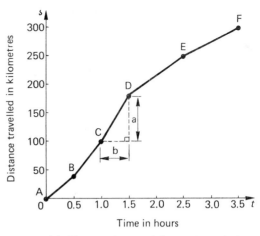

(a) Distance-time graph for sports car test

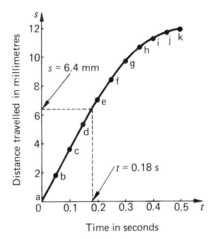

(b) Distance-time graph for a section of a vibrating spring

Fig. 5.1   Distance–time graphs for section 5.3 and calculations in section 5.5

axis and time on the horizontal axis) equals the speed at that point, i.e.

speed at a given point = gradient of distance–time graph at that point

For example, consider the simple distance–time graph of fig. 5.2(a) plotted for the case of a particle moving with constant speed. Since the graph is a straight line all points on the line AB have the same slope or gradient, so gradient of graph = speed of particle.

$$\text{gradient} = \text{slope of line AB}$$

$$= \frac{\text{BK}}{\text{AK}} = \frac{45\,\text{m}}{10\,\text{s}} = 4.5\,\text{m/s}$$

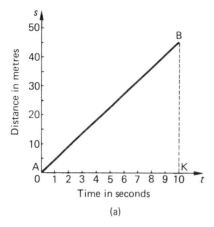

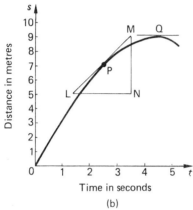

Fig. 5.2   Slope of curve at a point = speed at that point

but

$$\text{speed} = \frac{\text{distance travelled}}{\text{time taken}} = \frac{45}{10} = 4.5 \, \text{m/s},$$

so gradient of graph = speed of particle.

In fig. 5.2(b) we have plotted a distance–time graph where the speed of a body is changing with time and distance. Let us calculate the gradient at a point P on the graph; P corresponds to $t = 2.5 \, \text{s}, s = 7.1 \, \text{m}$. The gradient at P is found by first drawing a tangent to the curve at P (i.e. place a ruler so that it just touches the curve at P and draw the line LM (LM is the tangent)). Next, draw the horizontal line LN and the vertical line MN, so forming the right-angled triangle LMN:

gradient at P = slope of tangent to curve at P

$$= \frac{MN}{LN} = \frac{9-5}{3.5-1.4} = \frac{4}{2.1} = 1.9 \, \text{m/s}$$

= speed of body at P   (i.e. speed when $t = 2.5 \, \text{s}, s = 7.1 \, \text{m}$).

At point Q the tangent is parallel to the time axis and hence it has zero slope. Thus gradient at Q = speed of body at Q = 0 m/s.

Finally, for practice, plot the curve given in Table 5.2 and find the gradient and hence the speed of the body when (i) $t = 0 \, \text{s}$, (ii) $t = 0.2 \, \text{s}$, (iii) $s = 12 \, \text{mm}$. The answers are (i) 38 mm/s, (ii) 30.5 mm/s, (iii) 0 mm/s.

5.5
Calculations of average speed from given numerical and graphical data

Let us determine the average speed of the sports car whose distance–time data are given in Table 5.1, first over the individual time intervals, and then over the complete test run. The calculations, with the data repeated for convenience, are given below in Table 5.3. Note that over individual time intervals, the speed and average speed are the same since we have assumed constant speed in a given interval and neglected any time in acceleration or deceleration.

For the complete test run:

$$\text{total time} = 3.5 \, \text{h,}$$
$$\text{total distance travelled} = 300 \, \text{km,}$$

so average speed for total run

$$= \frac{\text{total distance}}{\text{total time}} = \frac{300}{3.5} = 85.7 \, \text{km/h}$$

$$= \frac{300 \times 1000 \, \text{m}}{3.5 \times 60 \times 60 \, \text{s}} = 23.8 \, \text{m/s.}$$

Note also that the speeds given in Table 5.3 may be calculated from the slopes of the straight-line sections of the distance–time graph plotted for the sport car test in fig. 5.1(a), e.g.

slope of section CD

$$= \frac{a}{b} = \frac{180-100}{1.5-1} = \frac{80}{0.5} = 160 \, \text{km/h.}$$

The average speed of the point on the spring, whose distance–time data are given in Table 5.2, over the quarter period $t = 0$ to 0.5 s, is

$$\text{average speed} = \frac{\text{total distance travelled}}{\text{total time}} = \frac{12}{0.5}$$

$$= 24 \, \text{mm/s} \quad \text{or} \quad 0.024 \, \text{m/s}$$

The average speed over a different time, for

Table 5.3   Calculation of average speed in a given time interval for sports car test

Time, t (h)	Distance, s (km)	Average speed $u = \dfrac{\text{distance travelled in interval}}{\text{time of interval}}$
0	0	
0.5	40	$u = \dfrac{40}{0.5} = 80\,\text{km/h}$ or $\dfrac{40 \times 1000}{0.5 \times 60 \times 60} = 22.22\,\text{m/s}$
1.0	100	$u = \dfrac{100 - 40}{1 - 0.5} = \dfrac{60}{0.5} = 120\,\text{km/h}$ or $33.33\,\text{m/s}$
1.5	180	$u = \dfrac{180 - 100}{1.5 - 1} = \dfrac{80}{0.5} = 160\,\text{km/h}$ or $44.44\,\text{m/s}$
2.5	250	$u = \dfrac{250 - 180}{2.5 - 1.5} = 70\,\text{km/h}$ or $19.44\,\text{m/s}$
3.5	300	$u = \dfrac{300 - 250}{3.5 - 2.5} = 50\,\text{km/h}$ or $13.89\,\text{m/s}$

example from $t = 0$ to $t = 0.18\,\text{s}$, may be obtained by first using the distance–time graph of fig. 5.1(b) to determine the distance travelled in this time: draw a vertical line through $t = 0.18\,\text{s}$ (shown dotted on graph) and find where it cuts the curve; read off distance on vertical axis corresponding to this point. From the graph we obtain, $s = 6.4\,\text{mm}$, so the average speed in the interval $t = 0$ to $t = 0.18\,\text{s}$ is

$$\text{average speed} = \frac{\text{distance travelled}}{\text{time taken}} = \frac{6.4}{0.18}$$

$$= 35.6\,\text{mm/s} \quad \text{or} \quad 0.0356\,\text{m/s}$$

Note that it is very important to understand the difference between average speed and instantaneous speed. The average speed in a given time interval of t seconds, say, when a distance of s metres has been traversed is given by

$$\bar{v} = \frac{s}{t}\,\text{m/s}$$

During this interval the actual speed at any given time, known as the instantaneous speed, is given by the gradient of the distance–time curve at the given time:

instantaneous speed v = gradient of distance–time curve at given time

Only when a body is travelling at a constant speed are the two equal.

5.6
The difference between speed and velocity

$$\text{speed} = \frac{\text{distance travelled}}{\text{time taken}}\,\text{m/s}$$

$$\text{velocity} = \frac{\text{distance travelled}}{\text{time taken}}\,\text{m/s in a defined direction}$$

Thus speed and velocity are defined identically as far as magnitude is concerned and have the same units, but speed takes no account of the direction of travel. Speed defines magnitude only. Velocity, however, defines the speed and the direction of the motion. Velocity is speed in a given direction. For example a car may be travelling due north at a speed of 80 km/h and another car may be travelling due east at 80 km/h. The speeds are identical at 80 km/h. Their velocities are different and would be defined as

80 km/h in the direction due N

80 km/h in the direction due E

Velocity must include direction as well as speed in its definition.

Quantities which specify magnitude only are known as scalar quantities, e.g. speed, mass, temperature, energy are scalar quantities. Quantities which have both magnitude and direction are known as vector quantities. Velocity is a vector quantity, speed is a scalar quantity.

5.7
The definition of acceleration

Acceleration is the rate of change of velocity with time.

Acceleration has the units of metres per second squared, m/s^2, and like velocity is a vector quantity and must be specified by both its magnitude and its direction.

If the velocity of a body travelling in a straight line at time t_1 is v_1 and the velocity of the body at a later time t_2 is v_2, then the acceleration of the body,

$$a = \frac{\text{change in velocity}}{\text{time taken}} = \frac{v_2 - v_1}{t_2 - t_1} \, m/s^2$$

If v_2 is greater than v_1, a is a positive acceleration; if v_2 is less than v_1 then a is a negative acceleration, also known as a deceleration. When v_1 changes at a constant rate to v_2, the acceleration is a constant one. If the rate of change of velocity is not constant, then a defined above is the average acceleration between t_1 and t_2.

Example 5.2
(a) Calculate the acceleration of a body which is moving in a straight line with a constant rate of change of velocity, and is observed at a time $t_1 = 5$ s to have a velocity $v_1 = 23$ m/s, and at time $t_2 = 8$ s to have a velocity $v_2 = 80$ m/s.
(b) A car has a speed of 15 m/s at the instant when its brakes are applied, and a speed of 3 m/s when its brakes have been in operation for 6 seconds. The motion of the car is in a straight line. Calculate the average acceleration during the interval the brakes are applied.

Solution
(a) Acceleration a

$$= \frac{v_2 - v_1}{t_2 - t_1} = \frac{80 - 23}{8 - 5} = \frac{57}{3} = 19 \, m/s^2$$

(b) Average acceleration

$$= \frac{\text{final speed} - \text{initial speed}}{\text{time brakes applied}}$$

$$= \frac{3 - 15}{6} = \frac{-12}{6} = -2 \, m/s^2$$

5.8
The calculation of the gradient of a velocity–time graph and the interpretation of this gradient as acceleration

The gradient at any point on a velocity–time graph, with velocity plotted vertically and time plotted horizontally, equals acceleration, i.e.

acceleration at a given point
= gradient of velocity–time graph at that point

The following data were obtained for three types of motion in a straight line. Table 5.4 refers to linear motion when the acceleration is constant; Table 5.5 refers to motion where the velocity is first constant and then undergoes constant deceleration; whilst Table 5.6 gives data characteristics of oscillatory or vibrational motion where the velocity is continually changing.

Table 5.4 Speed–time data when acceleration constant

Time, t (s)		Speed, v (m/s)
A	0	5
B	1	10
C	2	15
D	3	20
E	4	25
F	5	30

Table 5.5 Speed–time data for constant speed followed by constant deceleration

Time, t (s)		Speed, v (m/s)
a	0	30
b	1	30
c	2	30
d	2	20
e	4	10
f	5	0

Table 5.6 Speed–time data for vibrational motion in a straight line

Time, t (s)		Velocity or speed v (m/s)	Time, t (s)		Velocity or speed v (m/s)
a	0	10.0	f	0.25	0
b	0.05	9.5	g	0.3	−3.1
c	0.1	8.1	h	0.35	−5.9
d	0.15	5.9	j	0.4	−8.1
e	0.2	3.1	k	0.45	−9.5
Time, t (s)		Velocity or speed v (m/s)	Time, t (s)		Velocity or speed v (m/s)
l	0.5	−1	q	0.75	0
m	0.55	−9.5	r	0.8	3.1
n	0.6	−8.1	s	0.85	5.9
o	0.65	−5.9	t	0.9	8.1
p	0.7	−3.1	u	1.0	10.0

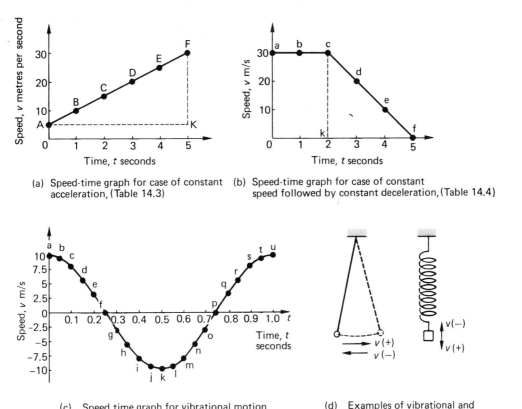

(a) Speed-time graph for case of constant acceleration, (Table 14.3)

(b) Speed-time graph for case of constant speed followed by constant deceleration, (Table 14.4)

(c) Speed time graph for vibrational motion, (Table 14.5)

(d) Examples of vibrational and oscillatory motion

Fig. 5.3 Velocity–time graphs. Consider the calculation of the gradient at various points on these velocity–time graphs:

$$gradient\ of\ velocity–time\ graph = acceleration$$

The corresponding velocity–time graphs are plotted in fig. 5.3. In each case, time is plotted along the horizontal axis and velocity along the vertical axis. The individual velocity–time points are first plotted. In graph (a) we join up the points A, B, C, D, E, F by a straight line—firstly since they lie in a straight line, and secondly because we are also told the motion has constant acceleration. This means that the velocity increases linearly with time. In graph (b) we join up the points a, b, c by a horizontal line, since from $t = 0$ to $t = 2$ s the velocity is constant. We then join up points c, d, e, f by a sloping straight line. This line indicates constant deceleration. In graph (c) we must draw a smooth curve through the plotted points (not join them up by straight lines), since the acceleration in this case is continually varying. Note, also, that we have both positive and negative values of velocity. This means that the motion actually changes direction, i.e. we have forward and backward

motion, typical of a vibrational type of motion. (d) shows two examples of linear motion which would have a velocity–time curve of the form shown in (c).

Graph (a)

Since the graph is a straight line, all points on the line AF have the same gradient, i.e.

$$gradient = \frac{FK}{AK} = \frac{(30 - 5)\,m/s}{5\,s} = \frac{25}{5} = 5\,m/s^2$$

but acceleration

$$= \frac{change\ in\ velocity}{time\ taken} = \frac{30 - 5}{5} = 5\,m/s^2,$$

so gradient = acceleration = $5\,m/s^2$.

Graph (b)

Over section abc, gradient = 0 (since line abc is

parallel to the time axis), hence gradient $=$ acceleration $= 0\,\text{m/s}^2$ from $t = 0$ to $t = 2\,\text{s}$.

Over section cdef,

$$\text{gradient} = \frac{-\text{ck}}{\text{fk}} = \frac{-30}{3} = -10\,\text{m/s}^2.$$

but the acceleration over this time is

$$\frac{\text{change in velocity}}{\text{time taken}} = \frac{\text{final} - \text{initial velocity}}{\text{time taken}}$$

$$= \frac{0 - 30}{5 - 2} = \frac{-30}{3} = -10\,\text{m/s}^2$$

so we have gradient $=$ acceleration (as, of course, we should expect). The minus sign denotes that a deceleration occurs, i.e. velocity decreases as time increases. Note also that the slope of a line or a tangent to a curve is taken as negative if the line makes an obtuse angle (angle between $90°$ and $180°$) with the horizontal axis. Thus the gradient of line cdef is negative, whilst the gradient of AF in graph (a) is positive.

Graph (c)
Redraw this graph on a larger scale (the data are given in Table 5.6), e.g. for velocity axis: 1 unit $=$ $1\,\text{m/s}$; for time axis, 1 unit $= 0.05\,\text{s}$. Draw the tangents to the curve at (i) $t = 0.15\,\text{s}$, (ii) $t = 0.4\,\text{s}$, (iii) $t = 0.75\,\text{s}$ and hence calculate the gradient of the velocity–time graph at these points. These give the acceleration at these points. The answers are (i) $-50.8\,\text{m/s}^2$, (ii) $-36.9\,\text{m/s}^2$, (iii) $62.8\,\text{m/s}^2$.

5.9
The solution of problems using the equation $s =$ (average velocity × time)

Remember, average speed or velocity

$$= \frac{\text{total distance travelled}}{\text{time taken}}$$

Thus, if we denote distance by s, time by t, and average velocity by $\bar{v}$, we have

$$\bar{v} = \frac{s}{t}, \quad s = \bar{v}t, \quad t = \frac{s}{\bar{v}}$$

For bodies which experience constant accele-

ration and travel in a straight line, average velocity, $\bar{v} = \frac{1}{2}(\text{initial velocity} + \text{final velocity})$.

Example 5.3
(a) A body is dropped from a high mast with zero initial velocity. Its velocity on striking the ground $3.8\,\text{s}$ later is $37.2\,\text{m/s}$. Calculate the height of the mast.
(b) The average velocity of an aircraft flying from A to B, a distance of $4000\,\text{km}$, is $800\,\text{km/h}$, whilst on the return trip from B to A the aircraft is aided by favourable winds which increase the average velocity on the return flight to $836\,\text{km/h}$. Assuming $0.5\,\text{h}$ is required for refuelling at B, calculate the time for the complete two-way journey.
(c) A body with an initial velocity of $6\,\text{m/s}$ rolls down an incline with constant acceleration. If the length of the incline is $120\,\text{m}$, and the time taken is $8\,\text{s}$, calculate the average and final velocities of the body.

Solution
(a) Average velocity during 'free-fall',

$$\bar{v} = \tfrac{1}{2}(\text{initial velocity} + \text{final velocity})$$
$$= \tfrac{1}{2}(0 + 37.2) = 18.6\,\text{m/s},$$

so the distance travelled,

$$s = \bar{v}t = 18.6 \times 3.8 = 70.7\,\text{m} = \text{height of mast}$$

(b) Time of flight A to B

$$= \frac{\text{distance travelled}}{\text{average velocity}} = \frac{4000}{800} = 5\,\text{h},$$

$$\text{time B to A} = \frac{4000}{836} = 4.785\,\text{h},$$

so the total time for the return journey, including refuelling, is

$$5 + 4.785 + 0.5 = 10.285\,\text{h}.$$

(c) Average velocity,

$$\bar{v} = \frac{\text{length of incline}}{\text{time taken}} = \frac{120}{8} = 15\,\text{m/s},$$

but

$$\bar{v} = \tfrac{1}{2}(\text{initial velocity} + \text{final velocity}),$$

so

$$15 = \tfrac{1}{2}(6 + v) \quad \text{where} \quad v = \text{final velocity},$$

i.e.

$$15 \times 2 = 6 + v, \quad v = 30 - 6 = 24\,\text{m/s}.$$

5.10
The determination of total distance travelled from velocity–time graphs: area under the curve of a velocity–time graph = distance travelled

The graph of velocity versus time for a body may be used to determine the distance travelled by the body. If we take the vertical axis for velocity v and the horizontal axis for time t, we may determine the distance travelled by calculating the area contained under the v–t curve and the horizontal axis:

area under v–t curve = total distance travelled in given time interval

Note that the proper use of units must be maintained, i.e. with v in metres per second, t in seconds, area under v–t curve = distance in metres.

The v–t graph drawn in fig. 5.4(a) shows an example where the velocity is constant at

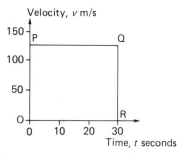

Velocity, v m/s

(a) Distance travelled in interval
$t = 0$ to 30 s = area OPQR
= OP × OR
= $125 \times 30 = 3750$ m

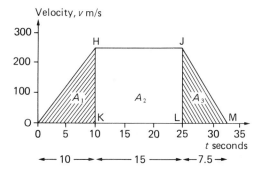

Velocity, v m/s

(b) Distance travelled = area OHJM
= $A_1 + A_2 + A_3$
where $A_1 = \tfrac{1}{2} \times 10 \times 250 = 1250$ m
$A_2 = 15 \times 250 \quad\;\; = 3750$ m
$A_3 = \tfrac{1}{2} \times 7.5 \times 250 = 937.5$ m
so $A \quad = 5937.5$ m

Fig. 5.4 Illustration of distance = area under v–t graph

$v = 125$ m/s, so the distance travelled in the interval $t = 0$ to $t = 30$ s is

$$s = \text{average velocity} \times \text{time}$$
$$= 125 \times 30 = 3750 \text{ m}$$

But the area under the v–t graph is

$$\text{area OPQR} = \text{OP} \times \text{OR}$$
$$= 125 \times 30 = 3750 \text{ m}$$

showing, for this simple case, distance = area under v–t curve.

In fig. 5.4(b), the velocity varies, increasing at a constant rate for the first 10 s, then remaining constant for 15 s and finally decreasing at a constant rate to zero after a further 7.5 s:

$$\text{total distance travelled} = \text{area under } v\text{–}t \text{ graph}$$
$$= \text{area } A_1 + \text{area } A_2$$
$$+ \text{area } A_3$$

but

$$\text{area } A_1 = \text{area of triangle OHK}$$
$$= \tfrac{1}{2} \text{ base} \times \text{height}$$
$$= \tfrac{1}{2} \times 10 \times 250 = 1250 \text{ m}$$

$$\text{area } A_2 = \text{area rectangle KHJL}$$
$$= 15 \times 250 = 3750 \text{ m}$$

$$\text{area } A_3 = \text{area of triangle LJM}$$
$$= \tfrac{1}{2} \times 7.5 \times 250 = 937.5 \text{ m}$$

so

$$\text{total distance travelled} = 1250 + 3750 + 937.5$$
$$= 5937.5 \text{ m}$$

The plotting of a velocity–time graph to determine distance is particularly useful when the velocity of the body varies with time. For example, let us consider the determination of distance travelled for which the following experimental results were obtained:

Time, t s	Velocity, v m/s	Time, t s	Velocity, v m/s
0	100	32.5	121
5	138	35	73
10	171	37.5	45
15	192	40	27
20	200	42.5	16
25	200	45	10
30	200	50	0

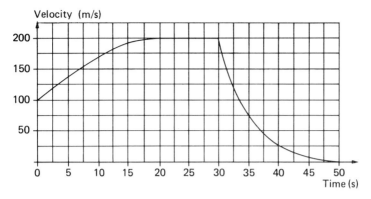

Fig. 5.5 Determination of distance by finding the area under the velocity–time graph. This may be accomplished by counting the number of 'squares' under the curve. Area of individual square = 2.5 s × 25 m/s = 62.5 m. Total number of squares under curve = 99.5, so work done = 99.5 × 62.5 ≈ 6200 m

We may determine the total distance travelled as follows:

(a) Plot the points given in the above table on graph paper, choosing suitable scales for the velocity and time axes.
(b) Draw a smooth curve through the plotted points.
(c) Calculate the area under the velocity–time curve. This may be accomplished by counting the number of 'squares' and fractions of 'squares' contained under the curve. For example, in the graph shown in fig. 5.5, each of the squares, taking into account the scale chosen, has dimensions:

$$2.5 \text{ s} \times 25 \text{ m/s} = 62.5 \text{ m}^2$$

i.e. each square corresponds to an area of 62.5 m².

(i) The number of squares under the curve in the 'acceleration' region between time $t = 0$ to 20 is approximately 52.5. Thus the distance travelled in this interval, is number of squares × area of individual square = 52.5 × 62.5 ≈ 3300 m.
(ii) The distance travelled in the interval $t = 20$ to 30 s where the velocity remains constant at 200 m/s may be easily calculated:

$$\text{distance} = 10 \text{ s} \times 200 \text{ m/s} = 2000 \text{ m}$$

We can, of course, count the number of squares: 32 in this region, so distance = 32 × 62.5 = 2000 m, which obviously checks with the above result.
(iii) The number of squares under the curve in the 'deceleration' region between $t = 30$ and 50 s is approximately 15. Thus the distance is

$$15 \times 62.5 \approx 940 \text{ m}$$

Thus the total distance travelled by the body

$$s = \text{total area under } v\text{–}t \text{ graph}$$
$$= 3300 + 2000 + 940 = 6240 \approx 6200 \text{ m}$$

We can of course, increase the accuracy of our result by choosing smaller 'squares', and thus increase the computed accuracy of the area under the curve.

5.11
The definition of force and its unit the newton (N)

A force may be taken as the measure of the effect which is capable of producing or affecting motion. In Chapter 2 we described the effect of forces in extending, compressing, and shearing materials. In Section 7.1 we define work done as equal to force × distance moved. In both cases we state that the SI unit of force is the *newton*, abbreviation N.

In fact the unit of force was named after Sir Isaac Newton, who first formulated the laws of motion. We still use these laws to solve the majority of problems concerning the motion of bodies.

Based on Newton's second law of motion, we define the force F acting on a body of mass m kilograms which has an acceleration of a metres per second squared as a result of the application of the force, as

$$F = ma \text{ newtons}$$

The SI unit of force, that is one newton (1 N), is that force which will accelerate a mass of one kilogram at the rate of one metre per second every second.

5.12
Acceleration is the result of a net force being applied

The acceleration of a body is due to the resultant or net force being applied to it.

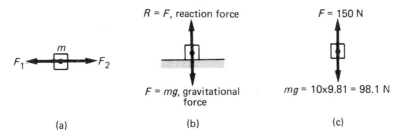

Fig. 5.6

For example, in fig. 5.6(a) a body of mass m kilograms is acted upon by two forces whose lines of action are in opposite directions. If F_2 is greater than F_1, the net force on the body is $(F_2 - F_1)$ newtons and the acceleration of the body in the direction left to right is found from

$$F_2 - F_1 = ma, \quad \text{i.e. } a = \frac{F_2 - F_1}{m} \, \text{m/s}^2$$

When a body is 'free-falling' under gravity the net force acting on the body is the gravitation force due to attraction between the body and earth minus the force of resistance due to the body's motion through the air and a small buoyancy force due to the body displacing a volume, equal to the body's volume, of air. If we neglect the latter two forces, the body accelerates at $g = 9.81 \, \text{m/s}^2$ and if the mass of the body is m kilograms, the gravitation force is

$$F = mg \text{ newtons}$$

In practice, however, a body undergoing 'free-fall' first accelerates with an acceleration close to g. As its velocity increases the air resistance to the body's motion also increases and so the net force on the body, and correspondingly its acceleration, is reduced. The body eventually reaches a velocity, known as the terminal velocity, beyond which any increase is not possible. At this stage the net force on the body is zero, the gravitational force being exactly counterbalanced by resistive forces and to a very small extent buoyancy forces. The terminal velocity of moderately sized raindrops is about 8 m/s; the terminal velocity of a parachutist in a delayed-opening jump may be as high as 250 km/h although his landing speed with parachute open is about 30 km/h.

Likewise, if a body is placed for example on a table as in fig. 5.6(b), the body is acted upon by a gravitational force of mg newtons, but the table provides an equal and opposite force, known as a reaction force, which exactly cancels the gravitational force. Hence there is zero net force on the body and obviously zero motion and acceleration. In fig. 5.6(c) a vertical force of $F = 150 \, \text{N}$ is applied to lift a mass of 10 kg; the net force acting on the mass is

$$F - mg = 150 - 10 \times 9.81 = 51.9 \, \text{N}$$

but net force $= ma$, so the acceleration of the mass is

$$a = \frac{51.9}{10} = 5.19 \, \text{m/s}^2$$

5.13
Motion under gravity problems (neglecting resistance) and application of the general formula $v = u + at$

A body dropped from a height above the earth's surface has constant acceleration. The gravitational force of attraction between the earth and the body produces a constant force on the body, which imparts to the body a constant acceleration. If we neglect the effects of air resistance, any body, regardless of its mass and shape, which is undergoing such a 'free-fall' has its velocity increased by 9.81 m/s for each second of its fall. Thus, if a body is dropped from rest, its velocity is 9.81 m/s after the first second, $2 \times 9.81 = 19.62$ m/s after the second second, and so on. We can thus describe the 'free-fall' of the body as an example of a body undergoing constant acceleration. In general, the velocity of a body after t seconds of 'free-fall' under gravity would be given by $v = gt$ m/s, where g is the constant acceleration due to gravity.

The numerical value of g varies slightly from point to point on the earth's surface; in fact, only $\pm 0.4\%$ about its average value. The official international standard of g is $9.80665 \, \text{m/s}^2$ measured at $45°$ latitude and at sea level.

If a body were not dropped from rest, but projected downwards with a velocity of u m/s, the velocity after t seconds would be given by $v = u + gt$. Similarly, if a body were projected vertically upwards with a velocity u, its velocity after t seconds would be $v = u - gt$, since in this case the constant acceleration due to gravity reduces the velocity on its upward path. Subsequently the body eventually comes to rest and will travel on a downward path with a constant acceleration of g m/s^2.

Any body acted upon by a constant force will in fact experience constant acceleration. The general formula for the velocity of the body's motion when undergoing constant acceleration is

$$v = u + at$$

where u = velocity in m/s at $t = 0$ (u is often known as the initial velocity),
a = acceleration in m/s^2, and
t = time in s.

Example 5.4
(a) A stone is dropped from rest from the top of a cliff. The stone strikes the water 1.8 s later. Assuming that the stone falls vertically and neglecting air resistance, calculate the velocity at which the stone strikes the water and the height of the cliff (take $g = 9.81$ m/s^2).
(b) A ball is projected vertically upwards with an initial velocity of 20 m/s. Calculate the time taken for the ball to reach its greatest height and the value of this height (take $g = 9.81$ m/s^2).
(c) A car starting from rest reaches 80 km/h in 10 s. Assuming that its acceleration is constant, calculate the acceleration and the distance travelled in the 10 s. If the brakes are then applied and produce a constant deceleration of 5 m/s^2, calculate the time taken for the car to come to rest.

Solution
(a) Applying $v = u + at$, with $u = 0$ and $a = g = 9.81$ m/s^2, we have

$$v = gt = 9.81 \times 1.8 = 17.7 \, \text{m/s}^2$$

$$= \text{velocity stone hits water.}$$

The average velocity $\bar{v} = \frac{1}{2}(u + v) = \frac{1}{2}(0 + 17.7) = 8.85$ m/s, so the distance travelled by the stone,

$$s = \bar{v}t = 8.85 \times 1.8 = 15.9 \, \text{m} = \text{height of cliff.}$$

(b) The velocity of the ball at a time t after the

instant of projection and on its upward path is

$$v = u + at = 20 - 9.81t,$$

since

$$u = 20 \, \text{m/s} \quad \text{and} \quad a = -g = -9.81 \, \text{m/s}^2.$$

When the ball reaches its greatest height its velocity is zero, if not it would travel a bit higher. The time taken to reach the greatest height can thus be calculated by setting $v = 0$, i.e.

$$v = 20 - 9.81t = 0 \quad \text{so} \quad t = \frac{20}{9.81} = 2.04 \, \text{s}$$

The average velocity of the upward motion,

$$\bar{v} = \frac{1}{2}(\text{initial velocity} + \text{final velocity})$$
$$= \frac{1}{2}(20 + 0) = 10 \, \text{m/s},$$

so the distance travelled, $s = \bar{v}t = 10 \times 2.04$ m = 20.4 m = greatest height.
(c) The velocity reached after 10 s,

$$v = 80 \, \text{km/h} = \frac{80 \times 1000}{60 \times 60} = 22.22 \, \text{m/s}$$

and thus applying $v = u + at$ with $u = 0$ (car initially at rest), we have

$$22.22 = 0 + a \, 10 \quad \text{so} \quad a = \frac{22.22}{10} = 2.22 \, \text{m/s}^2$$

The distance travelled in the 10 s is

$$s = \text{average speed} \times 10 = \frac{1}{2} \times (0 + v) \times 10$$
$$= \frac{1}{2} \times 22.22 \times 10 = 111.1 \, \text{m}$$

If the deceleration is constant and equal to 5 m/s^2 then $a = -5$ m/s^2, and since the car is braking to rest the final velocity is zero, so we have, on applying $v = u + at$, with $v = 0$, $u = 22.22$ m/s, $a = -5$ m/s^2, $0 = 22.22 - 5t$ so

$$t = \frac{22.22}{5} = 4.44 \, \text{s} = \text{time for car to come to rest}$$

5.14
Friction and friction forces

(a) The definition of friction force

Whenever a body is either in motion or tending to be forced to move there are always forces set up to oppose the motion or oppose the tendency of the body to move.

These forces are known as friction forces.

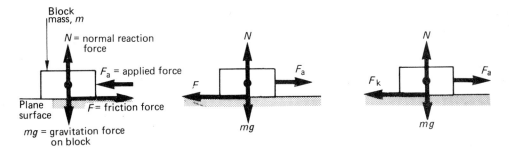

(a) Diagrams showing normal reaction N and friction force F
Note that F acts in opposite direction to applied force F_a

(b) Forces when block in motion
Friction force = F_k, the
kinetic friction force

Fig. 5.7

Friction forces oppose the relative motion between two surfaces, e.g. when a body is sliding on a surface its motion is impeded by friction forces acting at the surfaces of contact.

Friction forces also occur in liquids and gases; for example, when a body falls through air or when a boat moves through water friction forces oppose their motion.

Friction forces between solids are always present at the contacting surfaces of the solids when they slide or tend to slide over each other. The friction forces may be considered as the resistance to motion caused by surface irregularities. Even when the contacting surfaces appear smooth or even highly polished, a closer examination under a microscope will show the surface contours as consisting of a multitude of irregular valleys and peaks. Thus when an applied force acts to set up motion it will be opposed by the resistance produced by these irregularities tending to interlock.

We shall now define the friction force between two contacting solids with the aid of the diagrams shown in fig. 5.7, in which a block of mass m kilograms rests on a plane surface. The gravitation force on the mass is mg newtons acting vertically downwards. This force is exactly counterbalanced by the normal reaction force N of the plane on the block acting vertically upwards, so $N = mg$. N (or in general the perpendicular force pressing the two surfaces together) and the nature of the surfaces in contact are the primary factors determining the friction force. Next consider a small horizontal force F_a acting on the block but of insufficient magnitude to move the block. Immediately F_a is applied the friction force F is set up in the plane of contact to resist motion. The friction force $F = F_a$ and this equality remains so as F_a is

initially increased, but there comes a point at which the maximum friction force that can exist between the block and the plane is exceeded by F_a. At this point the block will begin to slide.

The friction force which prevents the block from moving is known as the static friction force. The maximum value of this force is known as the limiting friction force, which we shall denote by F_1. At the point when $F_a = F_1$ sliding is just about to take place. When F_a is greater than F_1 the block will move in the direction of the applied force. Actually, once sliding has been established it is found that the friction force opposing motion is slightly less than the limiting friction force. The friction force opposing motion when motion has been established is known as the kinetic friction force, F_k.

(b) The factors affecting the magnitude and direction of friction forces

The friction force always opposes the motion of a body and therefore always acts in the direction opposite to that of the motion and the applied force, as illustrated in fig. 5.7.

The magnitude of the friction force between two solids in contact and whose areas of contact are clean and dry, i.e. free from water, grease, oil, etc., depends on the following factors:

(1) The perpendicular force between the two surfaces of contact, e.g. the normal reaction force N in fig. 5.7.
(2) The nature of the materials in contact and their smoothness.
(3) The friction force does not normally depend on the actual area of contact, provided that this area is not so small as to cause penetration (e.g.

grooving) or seizing which would prevent normal sliding motion.

(4) The friction force is normally independent (at least for certain materials and over certain ranges of sliding speeds) of sliding speed, provided that any heat produced by sliding does not change surface conditions.

(5) The limiting friction force (maximum force of friction just before sliding occurs) is greater than the kinetic friction force (friction force when motion is established).

For clean dry surfaces it is possible to form a simple relationship between the kinetic friction force F_k and the perpendicular force N between two given surfaces in uniform sliding motion. The relationship is only approximate and does not apply under all conditions; for example it is inaccurate when N is relatively small and also when N is relatively high. The relationship, based on experimental observations, is

$$F_k = \mu_k N$$

where μ_k (Greek letter mu) is a constant for two given materials. μ is known as the coefficient of sliding or kinetic friction:

$$\mu_k = \frac{F_k}{N} = \frac{\text{kinetic friction force}}{\substack{\text{perpendicular or normal force} \\ \text{between the 2 surfaces}}}$$

A similar relationship can also be used to relate limiting friction F_1 and N:

$$F_1 = \mu_s N, \quad \mu_s = \frac{\text{limiting friction, } F_1}{\text{normal force, } N}$$

μ_s is known as the coefficient of static or limiting friction, and since F_1 is greater than F_k, μ_s is greater than μ_k. In practice μ_s is anything from a few percent to as much as 80 % higher than μ_k. Some typical values of the coefficient of kinetic friction are given in Table 5.7.

Table 5.7   Some average values of coefficients of sliding or kinetic friction for clean dry surfaces (except for lubricated steel)

Materials	μ_k
Wood on wood	0.4
Steel on steel	0.2
Steel on ice	0.04
Steel on concrete	0.3
Steel on steel (lubricated)	0.03
Rubber on concrete	0.75

(c) Examples of (1) practical applications, and (2) design implications of friction forces

(1) Practical applications of friction forces
Everyday examples of the application of friction forces are easy to find: when we strike a match the work done against friction is dissipated as heat and this raises the temperature of the match to ignition point and it bursts into flame; nails, screws, and nuts and bolts rely on friction forces to prevent them from working loose; when we apply the brakes in a car (see fig. 5.8(a)) the brake shoes are pressed firmly against the brake drum and the mechanical energy of motion is transformed into heat by the friction forces at the interface between the brake linings and inner drum surface; friction forces are used to transfer mechanical energy to a machine by means of a belt drive or friction clutch (*see* fig. 5.8(b)).

(2) Design implications of friction forces
Friction forces between moving surfaces of machines cause wear and the wasteful dissipation of energy as heat. Friction eventually causes either excessive wear or the surfaces to become permanently damaged, and the associated heat often causes additional damage to the surfaces as well as presenting the problem of providing adequate cooling to prevent overheating.

Thus it is essential in design to reduce friction effects. This will save energy, lengthen machine life, reduce maintenance, and generally lead to improved performance. Friction may be vastly reduced by lubricating contacting metal surfaces by oil or grease, as shown for example in Table 5.7, which shows that by lubrication the coefficient of kinetic friction between steel surfaces may be reduced from 0.2 to 0.03. In the ideal situation the contacting boundary surface is covered by a thin film of lubricant. This film reduces friction forces compared to 'dry-clean' friction by factors of over 10 to even 100 or more. The friction force for lubricated surfaces is also largely independent of the perpendicular force between the surfaces, but increases with surface area and with sliding speed.

Example 5.5
A body of mass 100 kg rests on a horizontal surface. The coefficients of static and kinetic friction between the two contacting surfaces are 0.25 and 0.2, respectively. Calculate first the minimum force to move the body, and then the force required to keep the body sliding at a

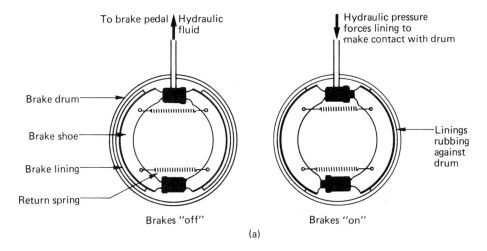

Brake drum

Brake shoe

Brake lining

Return spring

To brake pedal / Hydraulic fluid

Hydraulic pressure forces lining to make contact with drum

Linings rubbing against drum

Brakes "off"   Brakes "on"

(a)

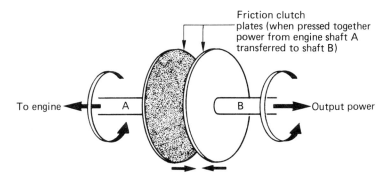

Friction clutch plates (when pressed together power from engine shaft A transferred to shaft B)

To engine ← A   B → Output power

(b)   Basic principle of friction clutch

Fig. 5.8

constant speed. If this speed were 1.5 m/s, calculate the power required. Take $g = 9.81$ m/s^2.

Solution
The body will start to move when the applied force is just about to exceed the limiting friction force,

$$F_1 = \mu_s N,$$

where μ_s = coefficient of static friction = 0.25,
    N = normal reaction between body and surface = 100 g.

Thus minimum force required to move the body is

$$F_a = F_1 = \mu_s N = 0.25 \times 100 \times 9.81 = 254.3 \text{ N}$$

Once the mass is in motion, the force F_a' required to maintain a constant sliding speed is equal to kinetic friction force, so

$$F_a' = F_k = \mu_k N = 0.2 \times 100 \times 9.81 = 196.2 \text{ N}$$

and the power,

$$P = \text{work done per second}$$
$$= \text{force} \times \text{distance moved per second}$$
$$= \text{force} \times \text{speed} = 196.2 \times 1.5 = 294.3 \text{ W}$$

Example 5.6
(a) Figure 5.9(a) shows a diagram of a steel block of mass 10 kg resting on the surface of a magnetic chuck. The coefficient of static friction between the block and the chuck is 0.2. The chuck can be magnetised and thus increase the normal force (N) between the block and chuck. If the block is subjected to a horizontal force of 80 N, calculate the minimum magnetising force which must be provided by the chuck to hold the block stationary. Take $g = 9.81$ m/s^2.
(b) The apparatus shown in fig. 5.9(b) was used to measure the coefficient of kinetic friction between the block of mass M kilograms and the plane surface. With masses $m_1 = 2$ kg and then

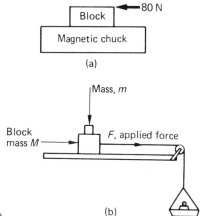

Fig. 5.9

$m_2 = 4$ kg placed on the block constant sliding speed was maintained with applied forces of $F_1 = 12$ N and $F_2 = 16$ N, respectively. Calculate the mass M of the block and the coefficient of kinetic friction, μ_k. Take $g = 9.81$ m/s^2.

Solution

(a) The condition that the block should not move is that the force of limiting friction must be equal to or greater than the applied force of 80 N. To achieve this we require the magnetisation force to at least increase the normal force by an amount N_m determined from

$$80 = \mu_s N = \mu_s(N_m + 10\,g)$$

as the normal force = magnetisation force + gravitation force due to block. Thus $80 = 0.2(N_m + 10 \times 9.81) = 0.2N_m + 49.05$, so $0.2N_m = 80 - 49$, $N_m = 31/0.2 = 155$ N ... minimum magnetising force.

(b) Using $F = \mu N$
where $N = (M + m)g$, the normal force due to block plus added mass, we have

$$F_1 = \mu_k(M + 2)g = 12 \qquad (1)$$
$$F_2 = \mu_k(M + 4)g = 16. \qquad (2)$$

On dividing (1) by (2) to cancel out μ_k we obtain,

$$\frac{M + 2}{M + 4} = \frac{12}{16} = \frac{3}{4}$$

so $4(M + 2) = 3(M + 4)$, i.e. $4M + 8 = 3M + 12$, so $M = 4$ kg.
Finally, on substituting $M = 4$ in (1), we have

$$\mu_k(4 + 2) \times 9.81 = 12,$$

i.e. $58.86\mu_k = 12$,   so $\mu_k = \dfrac{12}{58.86} = 0.20$

Problems 5:   Dynamics and friction

1.   Define speed and average speed. Calculate:
(a) The average speed of a body (falling under gravity) which travels 100 m in 4.5 s.
(b) The average speed of an athlete who runs 10 000 m in 28 minutes and 20 seconds.
(c) The time taken for sound to travel 2 km, given that the speed of sound is 331 m/s.
2.   The distance–time data for a body travelling at a varying speed in a straight line are given in the following table:

Time, t (seconds)	0	10	20	30	40	50
Distance, s (metres)	0	10	15	19	22	24

(a) Plot the distance–time graph for the body's motion (distance along vertical axis).
(b) Calculate the average speed in the time intervals $t = 10$ to 20 s and $t = 40$ to 50 s, and the overall average speed.
(c) Determine the value of the slope of the graph at $t = 10$ s, and interpret the meaning of this value.
(d) At what instants of time is the speed of the body at (i) a minimum, (ii) a maximum?
3.   Define acceleration. State the condition necessary for acceleration to occur. Calculate the average acceleration of:
(a) A body moving in a straight line which at time $t = 0$ has a velocity of 10 m/s, and at $t = 8$ s has a velocity of 34 m/s.
(b) A car braking in a straight line from a speed of 30 m/s to rest in 6 s.
   If in (b) the deceleration is constant, calculate the distance travelled by the car in braking from 30 m/s to rest.
4.   The velocity–time graph of fig. 5.10 describes the motion of a body moving in a straight line. Determine:
(a) The body's acceleration in the time intervals: $t = 0$ to 10 s, $t = 10$ to 20 s, $t = 20$ to 40 s.
(b) The distance travelled in the same three time intervals.
(c) The average velocity of the body over the complete time $t = 0$ to 40 s.
5.   Describe the characteristics of the motion of a body in a 'free-fall' under the action of gravity.
   A body is dropped from a high building with zero initial velocity and strikes the ground 3.2 s later. Calculate, taking $g = 9.81$ m/s^2:
(a) The velocity at which the body strikes the ground.

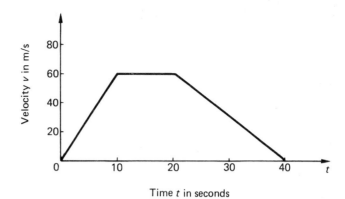

Fig. 5.10   Velocity–time   graph   for
             Problem 4

(b) The height of the building.

(c) The initial velocity that should be imparted to the body if, when projected upwards from the ground, it were to reach half-way up the building height.

6.   The equation of motion of a body experiencing constant acceleration is given by $v = u + at$. Define the meaning of the terms in the equation.

   A car travelling at an initial velocity of $10\,\text{m/s}$ accelerates at a constant rate and reaches a velocity of $40\,\text{m/s}$ in $5\,\text{s}$. Calculate (a) the acceleration and the distance travelled in the $5\,\text{s}$, and (b) the time taken for the car to come to rest and the distance travelled in coming to rest, if the brakes are subsequently applied and effect a constant deceleration of $8\,\text{m/s}^2$.

7   A body of mass $20\,\text{kg}$ is lifted vertically by a force of $300\,\text{N}$ acting for $4\,\text{s}$. Calculate, assuming $g = 9.81\,\text{m/s}^2$:

(a) The net force acting on the body during this time.

(b) The acceleration of the body.

(c) The distance moved in the $4\,\text{s}$ (assume the body is initially at rest).

   If at the end of the $4\,\text{s}$ the applied force is removed, calculate the maximum extra height that the body will reach.

8.   (a) A missile is projected vertically into the air and reaches its maximum height in $40\,\text{s}$. Determine the velocity of projection of the missile and the maximum height reached.

(b) A hot air balloon of total mass $1000\,\text{kg}$ rises vertically at the rate of $0.2\,\text{m/s}^2$. Calculate the total uplifting force which produces this acceleration. take $g = 9.81\,\text{m/s}^2$ and neglect the effects of resistance.

9.   (a) Define, with the aid of a diagram, the friction force.

(b) State the factors which determine the magnitude and direction of the friction force.

(c) Give two examples of (i) the practical application, and (ii) the design implication of friction forces.

10.   (a) Define the coefficient of kinetic or sliding friction.

(b) A body of mass $80\,\text{kg}$ rests on a horizontal surface. If the coefficient of kinetic friction between the two contacting surfaces is 0.2, calculate the force required to slide the body at a constant velocity. Take $g = 9.81\,\text{m/s}^2$.

(c) A machine is moved on skids across a horizontal floor at a steady speed of $1.5\,\text{m/s}$. If the mass of the machine plus skids is $600\,\text{kg}$ and the coefficient of kinetic friction is 0.15, determine the applied force to slide the machine. Take $g = 9.81\,\text{m/s}^2$.

6 Waves

The expected learning outcome of this chapter is that the student describes wave motion and solves problems involving wave velocity.

6.1
A list of simple examples of wave motion

The concepts of waves and wave motion are employed by scientists and engineers to describe and to predict the behaviour of many physical processes, all of which have in common the fact that energy may be transferred from one point to another in a medium (or, in the case of electromagnetic waves, also through a vacuum) without any net transportation of the medium accompanying the transfer of energy.

For example, when a wave of water moves from one place to another water does not move bodily with the wave. In fact, the water molecules have a circular or elliptical motion, as shown in fig. 6.1(a). When wind produces waves in a field of corn, the wave travels across the field, but the corn remains in place. In general, we may say that waves contain some form of energy and it is the wave motion that transfers this energy from point to point.

The learning objective of this sub-section is to be able to list examples of wave motion. Here are some common examples:

(1) *Surface water waves on oceans, lakes, rivers, canals, and ponds.* Certainly surface water waves are the most visual of all wave motions. Waves pass over the surface of water, carrying with them tremendous amounts of energy which, for example, is expended on our beaches, in breaking up rocks and eroding shore lines. Research is currently being done into the practical means of

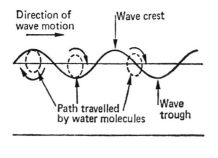

(a). Water surface waves

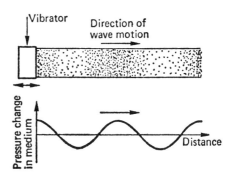

(b) Sound waves in a bar of material

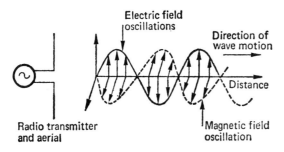

(c) Electromagnetic radio waves radiated from an aerial

Fig. 6.1

harnessing the wave energy of the seas around our coastline.

(2) *Sound waves in air, liquids, and solids.* Any source of vibration in a medium generates waves which propagate the energy of the source through the medium. Examples of various types of sound waves are: waves generated by speech and musical instruments, which propagate through air, taking with them the energy from the source to our ear; troublesome vibrations transmitted through buildings; seismic waves transmitted through the earth's crust by an earthquake; ultrasonic waves set up in water and used for depth sounding; ultrasonic waves propagated in a material or a component to test for faults within the material or component; noise generated by car and aero engines—the vibrations caused by the latter set up sound waves in air known as noise. A pictorial representation of a sound wave in a bar together with the associated pressure variation of the wave with distance is shown in fig. 6.1(b).

(3) *Waves on stretched strings.* Familiar examples are the waves set up on the strings of musical instruments or when a wire is plucked.

(4) *Electromagnetic waves.* X-rays, ultra-violet light, heat radiation, light, and radio waves are all examples of electromagnetic waves. These waves require no material medium for their transmission, in contrast to the waves stated above. They consist of variations of electric and magnetic fields which vibrate at right angles to the direction of the wave motion, as shown for example in fig. 6.1(c). These waves carry electromagnetic energy.

(5) *Probability waves.* Perhaps this is a rather obscure example but modern physicists and chemists apply wave concepts to predict the behaviour of electrons and atoms in their never-ending efforts to understand the properties and characteristics of matter and energy.

6.2
The explanation, using simple diagrams of (a) wavelength, and (b) frequency

(a) Wavelength

Wave motion may be demonstrated visually by vibrating one end of a loosely coiled spring—a coil of length 3 to 5 m and diameter about 50 mm is very suitable. In fig. 6.2(a) the vibrator alternately compresses (condenses) and extends (rarifies) the coil at end A, say at about 5 to 10 times per second.

These successive condensations and rarifications produced at end A travel down the coil and are absorbed at end B. We can observe that the distances between successive maxima of condensation, C_1 and C_2, or rarification, R_1 and R_2, are always the same, even though they are constantly moving from left to right. If we were to plot the relative packing of spring turns per centimetre versus the distance from end A at a given instant of time, we would obtain a graph of the form shown in (ii). At a slightly later instant of time, we would obtain a graph of exactly the same form but shifted along the distance axis as shown in (iii). We can think of (ii) and (iii) as 'photographs' taken at two relatively close but separate instants of time. We can use these wave curves to explain the meaning of wavelength.

The wavelength is the distance between any two successive points, in a medium in which a wave is propagating, which are experiencing identical changes, i.e. identical condensation or identical displacement, and the directions of change in these are the same, both either increasing or decreasing. Wavelength is denoted by the Greek letter lambda, λ. Thus, referring to the graph (ii) of compression versus distance;

$$\lambda = C_1C_2 = C_2C_3$$
$$= \text{distance between two maxima}$$
$$= R_1R_2 = R_2R_3$$
$$= \text{distance between two minima}$$
$$= Z_1Z_2 = Z_2Z_3$$
$$= \text{distance between two zeros}$$
$$= P_1P_2 \text{ (points where degrees of compression are identical)}$$

The wave motion simulated in fig. 6.2(a) is known as *longitudinal wave motion* since the spring vibrates in the same direction as the wave travels. In general, *longitudinal waves* are defined as waves which cause vibration of the medium in the direction in which the waves are travelling.

Transverse waves are waves whose displacements are at right angles to the direction at which waves are travelling. Vibrating strings, light waves and, to some extent, surface water waves are examples of transverse waves. Transverse waves are simulated in our long coil by vibrating end A at right angles to the coil axis. This situation is shown in fig. 6.2(b). The graph plotted is of transverse distance moved by the coil from its axis at a given instant of time. The

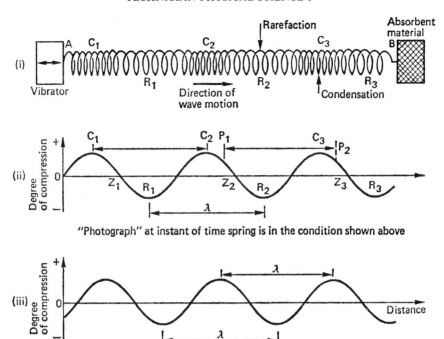

(a)   Longitudinal wave motion in a spring (analagous to sound waves)

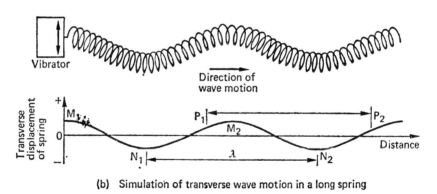

(b)   Simulation of transverse wave motion in a long spring

Fig. 6.2   Diagrams to explain the meaning of wavelength λ

wavelength for this wave is defined identically, i.e.

$$\lambda = M_1 M_2 = \text{distance between two maxima of transverse displacement}$$
$$= N_1 N_2 = \text{distance between two minima of transverse displacement}$$
$$= P_1 P_2 \text{ (distance between any two points with identical displacement and displacement change)}$$

Note that the wavelength of the longitudinal and transverse waves given above will not in general be equal.

(b) Frequency

Suppose we were to focus our attention on a particular section of spring and to plot the extent of condensation (i.e. degree of compression in the spring) as a function of time, that is, as time varies; we would obtain the variation shown in

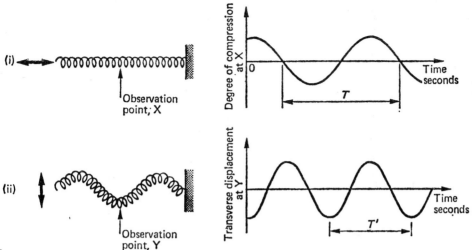

Fig. 6.3

fig. 6.3(i). This shows that there would be a complete cycle of events in which the section would undergo maximum condensation, return to zero condensation, and then be rarified and subsequently return to the unstressed (zero condensation) state, and so on.

The time taken for one complete cycle of the events at a given point is known as the periodic time T.

The number of cycles performed per second is known as the frequency of the wave. Thus, as one cycle takes T seconds, we have an important relation that the frequency:

$$f = 1/T$$

The frequency of the wave is identical with the frequency of vibrations produced by the source which creates the wave.

Finally, we can form an important relation between f and λ. In one period of T seconds, the wave travels λ metres. This must be so since at a given point we receive, for example, a maximum and T seconds later we receive another maximum. In this time, the first maximum has travelled in the direction of the wave, and in the identical time the second maximum has travelled from the direction of the source to our point of observation.

Thus the velocity of the wave:

$$v = \frac{\text{distance}}{\text{time}} = \frac{\lambda}{T} = \lambda \times \frac{1}{T}$$

but

$$\frac{1}{T} = f$$

so

$$v = f\lambda \text{ metres per second}$$

The SI unit of frequency is the hertz, abbreviation Hz. Common multiple units of frequency are:

1 kilohertz,	$1 \text{ kHz} = 10^3 \text{ Hz}$;
1 megahertz,	$1 \text{ MHz} = 10^6 \text{ Hz}$;
1 gigahertz,	$1 \text{ GHz} = 10^9 \text{ Hz}$.

The acoustic (sound) frequency range is normally taken as 0–20 kHz, although in practice we can only 'hear' sound from above 20 Hz or so up to frequencies of the order of 10 kHz to 16 kHz. Table 6.1 gives the classification of some of the main radio frequency ranges, together with their corresponding wavelength ranges, the latter calculated using

$$\lambda = \frac{c}{f} \text{ metres}$$

where $c = 3 \times 10^8$ m/s the velocity of light, $f =$ frequency in hertz.

The velocity of electromagnetic waves in free space $c = 3 \times 10^8$ m/s is constant. Electromagnetic waves travelling in dielectric media, however, travel slower. Their velocity is given by

$$v = \frac{c}{\sqrt{\varepsilon_r}}$$

where $\varepsilon_r =$ dielectric constant of the media in which the waves travel. For example, for polythene where $\varepsilon_r = 2.25$,

$$v = \frac{c}{\sqrt{\varepsilon_r}} = \frac{3 \times 10^8}{\sqrt{2.25}} = \frac{3 \times 10^8}{1.5} = 2 \times 10^8 \text{ m/s}$$

Table 6.1   Classification of radio frequency ranges

Frequency range	Wavelength range	Classification
30 kHz–300 kHz	10 000–1000 m	Low frequency (lf)
300 kHz–3 MHz	1000–100 m	Medium frequency (mf)
3 MHz–30 MHz	100–10 m	High frequency (hf) or short wave
30 MHz–300 MHz	10–1 m	Very high frequency (vhf)
300 MHz–3 GHz	1–0.1 m	Ultra high frequency (uhf)
1 GHz–300 GHz	0.3 m–1 mm	Microwaves

The velocity of sound also depends on the medium through which the waves travel. For example, the velocity of sound in air (at 0 °C) is 331.5 m/s and the velocity of sound in steel is 5100 m/s, which explains why we can 'hear' an approaching train first through the railway lines before we 'hear' it through the air. The velocity of sound depends on the medium and varies to some extent with temperature. It does not depend on the frequency or magnitude of the sound energy being propagated. Values of the velocity of sound through air, water, and some solid substances are given in Table 6.2.

Table 6.2   Velocity of sound in some common media

Medium	Velocity (m/s) (T in °C)
Air	$331.46 + 0.61T$
Water	$1403 + 4.2T - 0.028T^2$
Sea-water	$1449 + 3.6T$
Aluminium	5100
Copper	3600
Steel	5100
Wood	3000–4000
Glass	5000–6000
Brick	∼3700

6.3
The solution of simple problems using $v = f\lambda$

Remember:

$$v = f\lambda, \quad f = \frac{v}{\lambda}, \quad \lambda = \frac{v}{f}$$

where v = velocity of wave; units: metres per second, m/s
f = frequency (number of cycles or oscillations per second); units: hertz, Hz
λ = wavelength; units: metres, m

Example 6.1
(a) A diaphragm vibrating at a frequency of 512 Hz produces a sound wave in air of wavelength of 0.647 m. Calculate the velocity of sound in air.
(b) In an experiment to measure the velocity of electromagnetic waves in air the following results were obtained: wavelength $\lambda = 30.07$ mm, frequency of the source generating the waves, $f = 9.970$ GHz. Calculate the velocity of electromagnetic waves in air.
(c) In the definition of the metre it is stated that 1 650 763.73 wavelengths of the radiation emitted by the krypton atom equals one metre. Calculate the wavelength and frequency of this radiation to 4 significant figures, given that the velocity of light $c = 2.998 \times 10^8$ m/s.

Solution
(a) $f = 512$ Hz, $\lambda = 0.647$ m so the velocity of sound,

$$v = f\lambda = 512 \times 0.647 = 331.3 \text{ m/s}$$

(b) $\lambda = 30.07$ mm $= 30.07 \times 10^{-3}$ m, $f = 9.970$ GHz $= 9.970 \times 10^9$ Hz so the velocity of electromagnetic waves in air,

$$v = f\lambda = 30.07 \times 10^{-3} \times 9.970 \times 10^9$$
$$= 30.07 \times 9.970 \times 10^6 = 2.998 \times 10^8 \text{ m/s}$$

(c) 1 650 763.73 $\lambda = 1$ m, where λ = wavelength of radiation emitted by krypton atom, or approximately when working to 4 significant figures:

$$1.651 \times 10^6 \, \lambda = 1 \text{ m,}$$

so

$$\lambda = \frac{1}{1.651} \times 10^{-6} = 0.6057 \times 10^{-6} \text{ m}$$

The frequency of the radiation,

$$f = \frac{c}{\lambda} = \frac{2.998 \times 10^8}{0.6057 \times 10^{-6}}$$

$$= \frac{2.998}{0.6057} \times 10^{14} = 4.950 \times 10^{14} \text{ Hz}$$

Problems 6: Waves

1. Explain with reference to suitable diagrams the terms:
(a) wavelength and (b) frequency.
2. (a) Give four examples of wave motion.
(b) Draw graphs showing (i) the variation of pressure in a sound wave with distance at a given instant of time, (ii) the variation of pressure in a sound wave with time at a given distance from the sound source. Mark on the appropriate graph the wavelength and periodic time of the wave.

3. State the relationship between velocity, frequency and wavelength for waves.
A tuning fork vibrating at a frequency of 256 Hz produces a sound wave of wavelength 1.294 m. Calculate the velocity of sound in air.
4. A steel rail is vibrated at one end at a frequency of 1.5 kHz (1500 Hz). It is observed that the time for transmission of vibration through a 1 km length of rail is 0.196 s. Calculate the velocity and wavelength of the sound vibrations in steel.
5. (a) Calculate the frequency of a radio wave which has a wavelength of 1500 m.
(b) Calculate the wavelength of microwave radio transmissions at a frequency of 3 GHz (1 GHz = 10^9 Hz) propagating in free space.
The velocity of electromagnetic waves, $c = 3 \times 10^8$ m/s.

7 Work, energy, efficiency and power

The expected learning outcome of this chapter is that the student solves problems associated with energy.

Introduction: energy and work done

We all have some idea of what is meant by energy and work and the fact that work requires the expenditure of energy. Formally, we may relate energy and work by stating that energy is the capacity to do work.

In scientific terms, we shall be defining work as a force multiplied by the distance through which the force moves. Thus, to do work, a force must actually move through a distance. Some examples of doing work are: lifting a weight from a lower to a higher position—work is done against gravitational forces; to extend or compress a material or spring requires work to be done to overcome the elastic or cohesive forces of material or spring; a moving car is doing work against frictional forces, and if it moves up an incline it does work to overcome both these and gravity. If we were to push against an immovable object, such as a wall, we do no work in scientific terms, even though in human terms we may well feel that we are expending a considerable amount of energy.

To provide the necessary force and to do work by moving the force through a distance requires energy: when we lift a weight, energy is expended which enables our muscles to do the work; the actual energy supplied to our muscles comes indirectly from the food we eat. Most work in an industrial society is done by machines, whose energy is supplied in mechanical, electrical, and other forms, but whose basic source of input energy is normally via heat energy supplied by solid, liquid, or nuclear fuel.

Finally, to summarise: energy is the agency which enables work to be done. If our system or machine is perfect, all the energy supplied to the machine will be outputted as useful work. We shall see that energy has the same units as work, and that energy may be converted from one form to another so that work can be done.

7.1
The definition of work in terms of force applied and distance moved

Energy is expended by a source and work is said to be done on a body when a force overcomes the body's resistance to motion and moves the body through a distance. We define the work done W when a force F is applied to a body and the body moves in the direction of the line of action of the applied force by the product of the force and the distance s moved, i.e.

$$W = F \times s$$

In fig. 7.1(a) and (b) the work done by the applied force F in moving the body from position A to B, a distance s, is in both cases

$$W = F \times s$$

In case (b) the gravitational force acting on the body is equal to mg newtons, where m is the mass of the body in kilograms and $g = 9.81$ m/s^2 is the value of the acceleration due to gravity (*see*

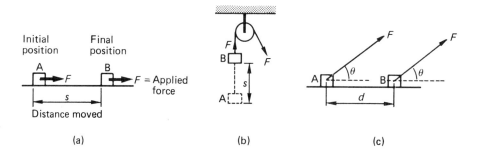

(a)   (b)   (c)

Fig. 7.1

Section 5.13). Thus, if we neglect frictional forces which may occur in the pulley system, the force F required and the work done W to move the body against gravity are given respectively by

$$F = mg, \quad W = mgs = 9.81 \; ms$$

If the direction of the force and the direction of motion are not in line, or not parallel, then the work done,

$W =$ (component of the force in direction of motion) × (distance moved)

Thus, in (c) the work done in moving the body from A to B is

$W =$ (component of force in direction of line AB) × d

$\quad = (F \cos \theta) \times d$

where $F \cos \theta$ is the component of F acting in the direction of line AB, θ being the angle F makes with the line AB.

The definition of the joule

The SI unit of work and energy is the joule, abbreviation J.

The joule is the amount of work which is done when a force of one newton moves a body one metre in the direction in which the force acts. Thus, one joule equals one newton metre, i.e. $J = Nm$.

Example 7.1

(a) Calculate the work done if a force of 50 N moves a body through 25 m in the direction of the force.
(b) Calculate the work done in lifting a mass of 60 kg through a vertical height of 70 m. Assume $g = 9.81$ m/s^2.
(c) A sledge is pulled along horizontal ground (at a uniform speed) by a rope which makes an angle of 25° with the horizontal. If the tension in the rope is $F = 90$ N, calculate the work done in moving the sledge 500 m.

Solution

(a) Work done = force × distance moved in direction of the force
$\quad = 50 \times 25 = 1250$ J

(b) The force required to lift a mass of $m = 60$ kg is

$$F = mg = 60 \times 9.81 = 588.6 \; N$$

and hence the work done in lifting the mass through 70 m is

$$W = 588.6 \times 70 = 41\,202 \; J$$

(c) The component of the force $F = 90$ N acting on the sledge in the horizontal direction of motion is

$$F \cos \theta = 90 \cos 25° = 90 \times 0.9063 = 81.57 \; N$$

hence the work done in moving the sledge 500 m is

$$W = 81.57 \times 500 = 40\,785 \; J$$

7.2
The drawing of graphs from experimental data of force–distance moved and the interpretation of work done as equal to the area under the graph

The body shown in fig. 7.2 is moved in three stages:

From A to B, distance moved $s_1 = 20$ m
applied force (parallel to AB), $F_1 = 90$ N

From B to C, distance moved $s_2 = 40$ m
applied force (parallel to BC), $F_2 = 60$ N

From C to D, distance moved $s_3 = 50$ m
applied force (parallel to CD), $F_3 = 34$ N

Our task is to draw a graph of force versus distance for the complete movement of the body from A to D. We select the vertical axis for force, 1 division equalling 20 N being a suitable scale, and use the horizontal axis for distance moved, 1 unit equalling 10 m being a suitable scale. The graph is drawn in fig. 7.2(b). Note that the force $F_1 = 90$ N from A to B is represented on the graph by line PQ, the force from B to C by line RS, and the force from C to D by the line TU.

Now the work done:

from A to B is
$\quad W_{AB} = F_1 \times s_1 = 90 \times 20 = 1800$ J
from B to C is
$\quad W_{BC} = F_2 \times s_2 = 60 \times 40 = 2400$ J
from C to D is
$\quad W_{CD} = F_3 \times s_3 = 34 \times 50 = 1700$ J

So that total work done in moving the body from point A to point D is given by

$$W = W_{AB} + W_{BC} + W_{CD}$$
$$= 1800 + 2400 + 1700 = 5900 \; J$$

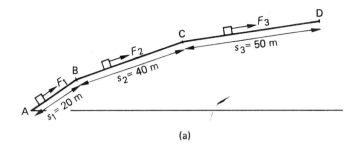

(a)

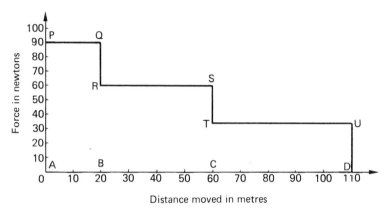

(b)   Force–distance moved graph for computing work done in
moving body from A to D

Fig. 7.2

But

$W_{AB} = PA \times AB = $ area PQBA
(i.e. area under graph of section PQ)

$W_{BC} = RB \times BC = $ area RSCB
(i.e. area under graph of section RS)

$W_{CD} = TC \times CD = $ TUDC
(i.e. area under graph of section TU)

so the total area under the graph = total work done, W.

This provides us with a very important result:

The area under a force–distance graph
= total work done

The plotting of a force–distance graph to determine the work done is particularly useful when the force varies its strength with distance. For example, let us consider the total work done in stretching a metal wire up to its limit of proportionality by plotting a force–distance graph. The following experimental results were obtained:

Force F(N)	0	20	40	60	80	100
Extension x(mm)	0	0.2	0.4	0.6	0.8	1.0

We may determine the total work done in extending the wire 1 mm by plotting the force–distance graph and finding the area under the graph. The graph is drawn in fig. 7.3. Note in this case, the distance is identical with extension, and in calculating the area we must convert the extension, although plotted in millimetres, to metres:

total work done = area under graph OG
$= $ area $\triangle$OGH
$= \frac{1}{2} \times $ OH $\times $ GH
$= \frac{1}{2} \times (1 \times 10^{-3}) \times 100 = 0.05$ J

7.3
Identification of different forms of energy and examples of energy conversions occurring in practical systems

(a) Different forms of energy

Energy can be regarded as consisting basically of two kinds: stored energy, which is known as potential energy, and energy due to motion, which is called kinetic energy. Potential and kinetic energy may be classified into various

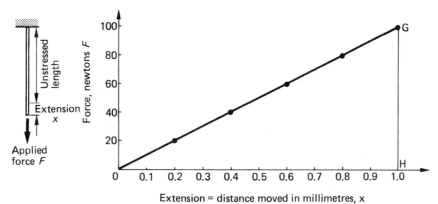

Fig. 7.3   Force–distance moved (extension) graph for a wire undergoing tensile stress. Total work done in complete extension of the wire equals area under graph

forms, depending on the source or origin from which the energy is drawn. We shall name and explain briefly six forms of energy. The specific learning objective of this sub-section is that you should be able to name at least five.

(1) Chemical energy
This is a form of potential energy stored in substances such as coal or petroleum fuels. When the fuel is burnt, the stored chemical energy is converted to heat energy. Chemical energy is also stored in primary and secondary cells (see Sections 11.4 and 11.5), which act as sources of electrical energy.

(2) Nuclear (atomic) energy
Nuclear energy is potential energy stored within the nuclei of atoms. In a nuclear reactor, a tiny quantity of matter is converted into energy by the fission or fusion of atomic nuclei. In the fission process, uranium nuclei are bombarded by tiny uncharged particles known as neutrons and energy is released because the nuclei so formed have less mass than the original uranium nuclei. The difference in mass m is converted to an amount of energy E according to Einstein's famous equation:

$$E = mc^2 \text{ joules}$$

where c = velocity of light = 3×10^8 m/s, m is in kilograms.

The energy released is mainly kinetic energy resulting from the tremendous speeds at which the nuclei formed in the fission process fly apart. If uncontrolled, this kinetic energy provides the destructive forces obtained in a nuclear explosion; when controlled, this energy is used to produce heat energy to generate steam. Subse-quently the steam may be used to drive a turbine, which in turn drives a generator to produce electrical energy or to provide mechanical energy to drive, for example, a nuclear submarine.

In the fusion process, two light nuclei, such as isotopes of hydrogen known as deuterium, combine to form helium, with a resulting loss in mass and hence a release of energy. Fusion processes account for the production of energy in our sun and the stars of the universe.

(3) Heat or thermal energy
Heat is a form of kinetic energy and is released in most chemical and all nuclear reactions. It is also generated when any work is done against frictional or resistive forces. It is exhibited as an energy source when heat is used to produce steam to provide sources of mechanical or electrical energy, or when petrol or diesel oil is ignited in a combustion engine.

(4) Mechanical energy
Mechanical energy is a form of kinetic energy. The energy contained in rotating flywheels and turbines provides examples of sources of mechanical energy. In an electrical generator, a coil is rotated in a magnetic field to produce a source of electrical energy; or vice versa in the electrical motor we pass current through the coil (which requires electrical energy to be supplied) and produce mechanical energy. Examples of the latter are electrical drilling machines and lathes.

(5) Electrical energy
In an electrical generating station, heat is converted to mechanical energy and this mechanical energy is used to produce electrical energy. A vast network of cables and lines (the

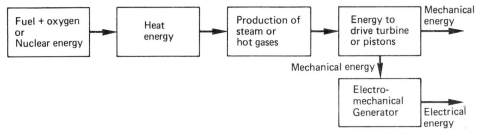

Fig. 7.4　Diagram shows processes involved in the conversion of heat energy to mechanical and electrical energy

grid system) transmits the electrical energy from the stations to industry and our homes. Electrical energy may also be obtained from batteries and accumulators by the conversion of the stored chemical energy in the latter. Electrical energy is perhaps the most practical and transportable of energies. We utilise electrical energy to drive machines, to provide heating and lighting, and to provide the power for our radios, televisions, etc.

(6) Light and radiation energy

The principal source of energy on our planet comes from electromagnetic radiation energy produced by the sun. The sun, in fact, generates this energy by nuclear fission. Electromagnetic energy is radiated out through space and it has been calculated that on average approximately 14 000 J of energy are received on the earth per square metre in each second.

The energy contained in the sun's radiation is a key factor in the growth of plant life, and hence in food production. Light energy was also directly responsible, although millions of years ago, for the production of the fuels of coal, oil, and gas (when plant and animal life decayed, they eventually formed these fuels).

Heat, light, and radio waves are forms of electromagnetic radiation. Light transmits energy, enabling us to see. Radio waves provide the means for the transmission of energy from point to point or over large areas, enabling us to use the telephone (over longer distances than immediately local calls, telephone messages are carried by radio waves transmitted via satellites or in coaxial cables buried in the ground or under the sea), listen to the radio, and watch television.

(b) Two examples of the conversion of heat energy to other forms of energy, and vice versa

(1) Examples of conversion of heat energy to mechanical energy and electrical energy

The block diagram of fig. 7.4 shows in simplified form the energy conversion process in the production of mechanical and electrical energy by the conversion of heat energy.

Initially, heat energy is released from the chemical energy stored in a fuel. The heat energy is then used to produce steam or heat gases (i.e. to increase the kinetic energy of water and/or gas molecules) to enable them to do work and drive, for example, a steam or gas turbine, or a piston in the case of an internal combustion engine. Thus, heat energy is converted to mechanical energy. The mechanical energy stored, for example, in a revolving turbine, may then be further converted in an electromechanical generator to produce electrical energy.

(2) Examples of the conversion of heat energy to chemical energy

A vast number of chemical reactions require heat energy before chemical combinations or physical changes can take place. For example, the solubility of many solid substances in water is increased by supplying heat energy. The heat energy initially provides energy to break down the substance into molecules or ions, and then further heat energy may be required in the subsequent process of the dispersion and interaction of these molecules with those of the liquid solvent. Heat energy must be initially supplied to ignite most fuels. After ignition, the chemical reaction of the fuel burning in oxygen releases heat energies in vast quantities.

(3) Examples of the conversion of mechanical energy to heat energy

All moving bodies and moving parts of machines experience frictional or similar resistive forces which restrict their motion. Thus, part of the mechanical energy of motion (i.e. kinetic energy) is always used to do work against these forces and is converted to heat. For example, the mechanical energy of a car in motion is reduced and converted to heat energy when the brakes are applied. The brake shoes or discs press against metal surfaces and work is done against the frictional forces at the interface of the brake and metal materials, heat being generated, and thus the mechanical energy contained in the motion of the car is reduced.

Heat is also generated by friction when a meteorite enters the earth's atmosphere. The kinetic energy of the meteorite is reduced at the expense of generating heat. The meteorite not only slows down very rapidly, but also burns itself away. Thus, it is extremely rare to find on the earth's surface a large meteorite. The development of special materials able to withstand the immense heat generation and associated high temperature involved when space vehicles re-enter the earth's atmosphere forms an important part of space research programmes in Russia and America.

(4) Examples of the conversion of electrical energy to heat

When electrical current flows in a conductor, work is done against the resistance presented by the conductor and electrical energy is converted to heat. Practical examples of the conversion of electrical energy to heat are the electric fire and the electric water heater.

As a further point to complete this sub-section, it is useful to remember that when energy is converted from one form to another, no energy is actually lost nor is any energy created. Thus, in all processes other than nuclear reactions, we can state that: energy cannot be created or destroyed. This statement is known as the principle of conservation of energy.

At one time, it was thought that mass could not be created or destroyed. However, with the advent of nuclear physics, we now know that mass can be converted to energy. This fact, together with the principle of conservation of energy, is now embodied in the principle of conservation of energy and mass, which states that:

the total energy and mass in any closed system is conserved, i.e. remains the same.

7.4
The definition of efficiency in terms of energy input and output

Any practical machine may be regarded essen-

tially as an energy converter. Energy of one kind is supplied and energy of other kinds (in some cases, of the same kind) is generated. In the process of energy conversion, some energy is inevitably wasted and not put to useful work. Machines have certain components, such as bearings, gears, and other moving parts which have to overcome frictional or other similar resistive forces in order to perform their function. Work done against these forces causes part of the energy supplied to the machine to be dissipated and wasted as heat. Consequently, the energy output of a machine is always less than the energy input to the machine, i.e.

energy output = energy input
　　　　　　　　 − energy losses due to work
　　　　　　　　　　 against resistive forces

The efficiency (normally denoted by the Greek letter eta η) is defined as

efficiency, η
$$= \frac{\text{energy output from machine or device}}{\text{energy input to machine or device}}$$

and the percentage efficiency,

$$\text{efficiency } \% = \frac{\text{energy output}}{\text{energy input}} \times 100\%$$

For a practical machine, efficiency is always less than 1, as illustrated in the following examples:

(1) If the energy input to a machine is 2000 J and the energy output is 1700 J, the efficiency of the machine is

$$\eta = \frac{\text{energy output}}{\text{energy input}} = \frac{1700}{2000} = 0.85, \text{ or } 85\%$$

(2) If a machine has an efficiency of 50% (i.e. $\eta = 0.5$) and an energy output of 4000 J is required, then as

$$\eta = \frac{\text{energy output}}{\text{energy input}}$$

$$\text{energy input} = \frac{\text{energy output}}{\eta} = \frac{4000}{0.5} = 8000 \text{ J}$$

7.5
Power is the rate of transfer of energy

In practice we are concerned not only with the total energy that may be supplied by a machine, but also with the rate of supply of this energy, since this dictates the time required to accomplish a specific task. That is, we require to know the rate at which energy is transferred from a machine to do work. We use the term 'power' to describe the time rate of energy transferred.

Power is defined as the rate at which energy is transferred to do work, i.e. power is the rate in joules per second at which energy is transferred or at which work is done.

Thus, if

> W = an amount of energy,
> t = the time in seconds in which this energy was transferred,

then, assuming a uniform transfer of energy with time, the power (the symbol P is normally used to denote power) is

$$P = \frac{W}{t}$$

If the transfer of energy varies with time, $P = W/t$ is the average or mean power.

The SI unit of power is the watt, abbreviation W

The watt equals the amount of power provided when one joule of energy is transferred in one second, i.e. 1 watt = 1 joule per second.

Useful multiple and sub-multiple units of power are:

> 1 megawatt (MW) = 10^6 watts,
> i.e. 1 MW = 10^6 W.
> 1 kilowatt (kW) = 10^3 watts,
> i.e. 1 kW = 10^3 W.
> 1 milliwatt (mW) = 0·001 watts,
> i.e. 1 mW = 10^{-3} W.

Example 7.2
The power input to a machine of 75 % efficiency is 800 W. Calculate the power output and the energy transferred in 20 minutes (i) to do work, (ii) to overcome losses in the machine.

Solution
Now power,

$$P = \frac{W}{t}$$

where W = energy or work done,
t = time in which energy is supplied or in which work is done.

Also efficiency,

$$\eta = \frac{\text{energy output}}{\text{energy input}} = \frac{W_o}{W_i}$$

where W_o = energy output = $P_o \times t$,
P_o = power output,
W_i = energy input = $P_i \times t$,
P_i = power input.

Thus

$$\eta = \frac{W_o}{W_1} = \frac{P_o t}{P_i t} = \frac{P_o}{P_i}$$

so

$$P_o = \eta P_i = 0.75 \times 800 = 600\,\text{W}$$

as

$$\eta = 0.75\ (75\,\%), \quad P_i = 800\ \text{W}.$$

The energy transferred in 20 minutes to do work is

$$W_o = P_o t = 600 \times (20 \times 60) = 720\,000\,\text{J}$$

(remember t must be in seconds, i.e. 20 minutes = 20×60 s).

The total energy supplied to the machine in 20 minutes is

$$W_i = P_i t = 800 \times (20 \times 60) = 960\,000\,\text{J}$$

so the energy dissipated in the machine itself to overcome losses is

$$W_i - W_o = 960\,000 - 720\,000 = 240\,000\,\text{J}$$

Alternatively the energy dissipated in the machine may be calculated from

$$(P_i - P_o)t = (800 - 600) \times (20 \times 60)$$
$$= 240\,000\,\text{J}$$

7.6
The solution of problems involving work, energy, efficiency and power

Before applying the results of the previous sections to some practical problems, let us first recapitulate on the important results.

(1) Work = force (in newtons)
$\qquad\qquad\qquad$ × distance moved (in metres)
i.e. $\quad W = F \times s$ joules.
(2) Area under a force–distance graph = work.
(3) Power = rate of transfer of energy
$\qquad\qquad\qquad\qquad$ or rate of doing work
i.e.
$$P = \frac{W}{t}$$

where W = work or energy in joules (J)
t = time in seconds (s)
P = power in watts (W = J/s)

(4) Efficiency $= \dfrac{\text{energy output}}{\text{energy input}}$ or $\dfrac{\text{power output}}{\text{power input}}$

Example 7.3

Calculate the work done and power required to lift a mass of 100 kg through a height of 40 m in 60 s. Take $g = 9.81$ m/s. If the power is supplied by an electric motor requiring a power input of 900 W, calculate the efficiency of the system.

Solution

The force required to lift the mass against gravity,

$$F = mg = 100 \times 9.81 = 981 \text{ N}$$

and hence the work done in lifting the mass through 40 m,

$$W = F \times s = 981 \times 40 = 39\,240 \text{ J}$$

The power required is the rate of doing work, i.e.

$$P = \frac{W}{t} = \frac{39\,240}{60} = 654 \text{ W}$$

The efficiency,

$$\eta = \frac{\text{power output}}{\text{power input}} = \frac{654}{900} = 0.727 \quad \text{or } 72.7\%$$

Example 7.4

A body is hauled up an inclined plane at a constant velocity of 1.5 m/s by a winching machine. To overcome gravity and the frictional force along the plane, the winch has to exert a force causing a tension of 2000 N in the hauling cable. Calculate the work done in moving the body 30 m up the plane and the useful power supplied by the winch.

Solution

Work done = force × distance moved
$= 2000 \times 30 = 60\,000$ J

The useful power supplied,

P = rate of doing work
= force × distance moved per second
= force × velocity = 2000×1.5
= 3000 W

Example 7.5

Figure 7.5 shows the force–distance graph for a

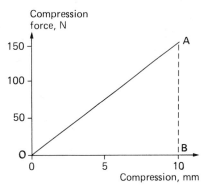

Fig. 7.5   Force–distance graph for spring of Example 7.5

spring which has undergone compression. Calculate the energy stored in the spring.

Solution

The energy transferred and stored in the spring

= work done in compressing spring
= area under force–distance graph
= area of triangle
= $\frac{1}{2} \times 150$ N $\times 10$ mm = 750 N mm
= 0.75 Nm = 0.75 J

Example 7.6

Figure 7.6 shows the force–distance graph for a load. Determine (a) the work done in moving the load 40 m, (b) the work done between 40 m and 80 m and (c) the total work done from 0 to 100 m. If the time taken to move the load 100 m is 200 s, determine also the average power supplied to the load.

Solution

Work done in moving the load,
(a) From 0 to 40 m,

$$W_1 = \text{area } A_1 = \tfrac{1}{2}(20 + 40) \times 40 = 1200 \text{ J};$$

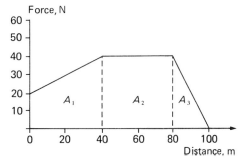

Fig. 7.6   Force–distance graph for Example 7.6

(b) From 40 to 80 m,

$W_2 = $ area $A_2 = 40 \times 40 = 1600$ J;

(c) From 0 to 100 m,

$W = $ area $A_1 + A_2 + A_3 = 2800 + \frac{1}{2} \times 40 \times 20$
$= 3200$ J.

Average power supplied, $P = W/t = 3200/200 = 16$ W.

Problems 7:   Work, energy, efficiency and power

1.   Define work, power, the joule, the watt. Describe the relationship between energy and work done. Calculate:
(a) The work done when a force of 150 N moves a body 200 m in the direction of the force.
(b) The work done in lifting a 80 kg mass through a vertical height of 44 m (take $g = 9.81$ m/s^2).
(c) The total energy supplied to a 2-kW electric fire switched on for 20 minutes.
2.   A force F increases linearly with distance s according to the formula: $F = 12s + 100$.

Complete the following table of values for s and F and plot a force versus distance graph. Hence calculate the total work done when the force moves from $s = 0$ to $s = 50$ m. Calculate also the work done in the interval from $s = 10$ m to $s = 30$ m.

Distance s (metres)	0	10	20	30	40	50
Force F (newtons)	100	220				

3.   Name five forms of energy and describe two examples of processes by which one form of energy may be converted to another.
4.   Define the term efficiency and explain why this is always less than 100% for a practical machine. Calculate:
(a) The efficiency of a machine with a power input of 5000 W and a power output of 4100 W.
(b) The energy input required to produce an output of 60 000 J from a machine of 60% efficiency.
5.   From the graph of force versus distance moved shown in fig. 7.7 calculate the work done in moving a load the distance 0 m to 6 m.

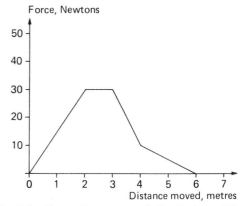

Fig. 7.7   Force–distance graph for Problem 7.5

8   Heat

The expected learning outcome of this chapter is that the student shows a basic understanding of heat and temperature.

Introduction: (a) The distinction between heat and temperature; (b) The Celsius temperature scale and the relationship between degrees celsius and kelvin

(a) *The distinction between heat and temperature*

Heat is a form of energy and may be converted into other forms of energy. Heat is measured in units of energy, that is, joules.

Temperature describes the degree of 'hotness' or 'coldness' of a body or medium. It is a means of specifying the 'level' of the sensation caused by heat energy, not the quantity of heat energy. For example, a lighted match can be at a much higher temperature than a full kettle of boiling water, but the latter contains considerably more heat energy.

Temperature may also be defined as a property which determines the direction of flow of heat energy. Heat energy always flows from a higher to a lower temperature (unless mechanical energy is expended to force it to do otherwise). For example, if a body at a higher temperature is immersed in water, heat from the body will be transferred to the water, with the result that the temperature of the water rises and the temperature of the body falls until the two temperatures are equal. If this occurs under insulated conditions so that no heat escapes to the surroundings, then equilibrium occurs when the heat loss by the body equals the heat gain by the water.

All matter possesses heat energy present in the kinetic energy (energy due to motion) of its atoms and molecules. When a body is heated, the motion of its molecules becomes more vigorous, thereby increasing their kinetic energy. The kinetic energy increases with temperature rise and the total kinetic energy of all molecules in a body is a measure of the heat energy contained in that body.

Finally, let us summarise how we differentiate between heat and temperature:

1. Heat is a form of energy, measured in joules.
2. Temperature is not an energy; it describes the 'level' but not the magnitude of heat, and dictates the direction in which heat will flow, i.e. from a higher to a lower temperature.

(b) *The Celsius temperature scale and the relationship between degrees celsius and kelvin*

In order to produce a temperature scale, a minimum of two fixed points is required to define the scale. These fixed points must always be very accurately reproducible. The two fixed points on the Celsius scale of temperature are:

(1) The lower fixed point, known as the ice point, which is defined as the temperature of melting ice, or, to be very specific, the temperature of equilibrium of ice, liquid water, and its vapour (this point is known as the triple point of water) at normal atmospheric pressure ($101\,325\,N/m^2$).
(2) The upper fixed point, known as the steam point, is defined as the temperature of steam rising from pure water boiling under normal atmospheric pressure.

On the Celsius scale, the ice point is taken as zero degrees celsius, and written as 0 °C,* and the steam point as one hundred degrees celsius, written as 100 °C. The temperature interval between the ice and steam points is divided into 100 equal divisions. Each division is called a degree and a temperature of 50 °C, for example, is half-way between the lower and upper fixed points.

Note that in everyday usage, the term degrees centigrade is used, but it is standard scientific practice to use degrees celsius and not centigrade. This is because of possible confusion with a new SI unit, the grade, which is one hundredth part of a right angle, a centigrade being therefore one ten-thousandth part of 90°.

The SI unit of temperature is the kelvin, abbreviation K. The kelvin is defined as 1/273.16

---

* To be absolutely accurate, the ice point is internationally agreed to be +0.01 °C, but for our purposes negligible error occurs in assuming the ice point to be 0 °C.

of the temperature at the triple point of water (the ice point). A temperature interval of one kelvin is exactly equal to a temperature interval of one degree celsius. The relationships between corresponding temperatures on the two scales are:

$$0\,\text{K} = -273.15\,°\text{C} \ldots \text{known as the absolute zero of temperature}$$
$$273.16\,\text{K} = +0.01\,°\text{C} \ldots \text{the ice point or triple point of water}$$
$$373.15\,\text{K} = 100\,°\text{C} \ldots \text{the steam point}$$

$$T\,\text{kelvin} = t\,\text{celsius} + 273.15$$

Thus, to convert a temperature quoted in kelvin to degrees celsius, add 273.15; usually 273 is accurate enough. To convert degrees celsius to kelvins, subtract 273.15. e.g.

$$20\,°\text{C} = 20 + 273 = 293\,\text{K}$$
$$-60\,°\text{C} = -60 + 273 = 213\,\text{K}$$
$$400\,\text{K} = 400 - 273 = 127\,°\text{C}$$
$$150\,\text{K} = 150 - 273 = -123\,°\text{C}$$

8.1
The definition of specific heat capacity

The specific heat capacity of a substance is the amount of heat energy required to change the temperature of one kilogram mass of the substance by one kelvin (or equivalently one degree celsius).

The units of specific heat capacity are joules per kilogram per kelvin, i.e. $\text{J kg}^{-1}\,\text{K}^{-1}$. Note that a temperature change of $1\,\text{K} = 1\,°\text{C}$ change, so we can also express the units as $\text{J kg}^{-1}\,°\text{C}^{-1}$. Specific heat capacities, like most physical quantities, e.g. Young's modulus, velocity of sound, and resistivity, depend on temperature. Values for some common substances over a given temperature range are given in Table 8.1.

8.2
The solution of problems associated with mass, specific heat capacity and temperature change

To solve these types of problem we must remember two facts:

(1) The amount of heat energy required to change the temperature of a mass of substance is given by

$$H = mcT \text{ joules}$$

Table 8.1 Specific heat capacity of some common substances

Substance	Specific heat capacity $(\text{J kg}^{-1}\text{K}^{-1})$	Temperature range $(°\text{C})$
Aluminium	908	17–100
Copper	385	15–100
Mercury	139.3	20
Iron	460	18–100
Steel	450–480	15–100
Sand	800	20–100
Stone	750–960	0–30
Ethyl alcohol	2290	0
Ice	2000–2090	−21–0
Water	4217	0
	4192	10
	4186	15
	4182	20
	4178	30–40
	4184	60
	4196	80
	4215	99

where m = mass of substance in kilograms,
c = specific heat capacity in $\text{J kg}^{-1}\,\text{K}^{-1}$ or $\text{J kg}^{-1}\,°\text{C}^{-1}$,
T = temperature change in kelvin or degrees celsius.

The amount of heat H is given out when the substance cools and is absorbed when the substance is heated.

(2) Total heat lost or transferred from a cooling body or bodies = total heat gained by or transferred to cooler body or bodies, e.g. if a body of mass m_1, specific heat capacity c_1 and at a temperature of $T_1\,°\text{C}$, is placed into a vessel of mass m_2, specific heat capacity c_2, containing a liquid of mass m_3, specific heat capacity c_3, and both are at a lower temperature T_2, then the final temperature $T\,°\text{C}$ of all three bodies may be found by applying 2, that is:

$$\text{Heat lost} = m_1 c_1 (T_1 - T)$$
$$= \text{heat gained}$$
$$= m_2 c_2 (T - T_2) + m_3 c_3 (T - T_2)$$

This equation may be solved to find T.

Example 8.1
Calculate the amount of heat energy required to

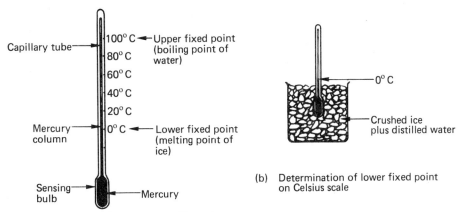

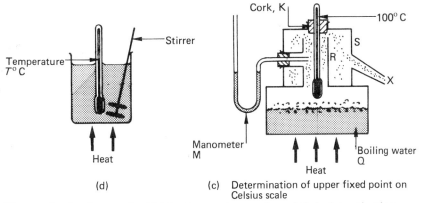

Fig. 8.1   The two fixed points on the Celsius temperature scale and their determination

raise the temperature of 200 kg of water contained in a copper tank of mass 20 kg from 20 °C
to 45 °C. The specific heat capacities of water and
copper are 4180 J kg^{-1} °C^{-1} and 385 J kg^{-1} °C^{-1}.

Solution
Heat required to raise water from 20 °C to 45 °C,

H_w = mass water × specific heat capacity
      × temperature rise
      = 200 × 4180 × (45 − 20) = 20.9 × 10^6 J

Heat required to raise copper tank through same
temperature change,

$$H_c = 20 \times 385 \times (45 - 20)$$
$$= 192\,500\,\text{J} = 0.1925 \times 10^6\,\text{J}$$

Total heat required = $H_w + H_c$ = 21·0925 × 10^6 J
$\simeq$ 21·09 MJ (megajoules).

Example 8.2
An 0.500 kg mass of aluminium is immersed in

water at 99 °C and reaches the temperature of this
water. The mass is then rapidly transferred and
immersed in 2 kg of water at 15 °C. Assuming no
heat losses (and no hot water is transferred with
the mass of aluminium), calculate the temperature to which the water rises. Take the specific
heat capacities of aluminium and water as 908
J kg^{-1} °C^{-1} and 4186 J kg^{-1} °C^{-1}.

Solution
Let the temperature to which water rises (and
aluminium falls) be T = C, then
heat transferred to water by aluminium cooling
from 99 °C to T °C = heat gained by water in
rising from 15 °C to T °C
so

$$0.5 \times 908(99 - T) = 2 \times 4186(T - 15)$$
$$44\,946 - 454T = 8372T - 125\,580$$
$$44\,946 + 125\,580 = 8372T + 454T$$

$$170\,526 = 8826T, \quad \text{so} \quad T = \frac{170\,526}{8826} = 19.3\,°C$$

Example 8.3

3 kW of power is supplied to heat a tank containing 150 kg of water initially at 20 °C. Neglecting energy losses, calculate the time the power must be kept switched on if the water is to have a final temperature of 50 °C. Specific heat capacity of water is 4180 J kg^{-1} °C^{-1}.

Solution

Total energy required to raise 150 kg from 20 °C to 50 °C,

$$W = 150 \times 4180 \times (50 - 20) = 18.81 \times 10^6 \text{ J}$$

Remembering that,

$$\text{energy (joules)} = \text{power (in watts)}$$
$$\times \text{ time (in seconds)}$$

we have for the time t seconds for which the power $P = 3 \text{ kW} = 3 \times 10^3 \text{ W}$ must be supplied:

$$t = \frac{W}{P} = \frac{18.81 \times 10^6}{3 \times 10^3}$$
$$= 6.27 \times 10^3 \text{ s (or 1.74 hours)}$$

8.3
The distinction between sensible heat and latent heat, and the form of temperature–time graphs for substances changing state

(a) The distinction between sensible heat and latent heat

The specific learning objective of this sub-section is that you should be able to differentiate between sensible heat and latent heat.

Sensible heat is that heat whose effect is observed as an increase or decrease in temperature. For example, if a substance is heated by an amount of energy H, its temperature rises from T_1 to T_2 according to the equation

$$H = mc(T_2 - T_1)$$

where m = mass of substance, c = its specific heat capacity. Likewise, if a substance cools from a temperature T_3 to a lower temperature T_4, the heat energy given out is

$$H' = mc(T_3 - T_4)$$

In both cases, a temperature change is observed and H and H' are sensible heats.

When a substance changes state, that is, it changes from solid to liquid form, or from liquid to gaseous or vapour form, heat energy must be supplied to effect the change. Whilst the change is taking place, there is normally no change of temperature. All the heat energy is used to effect the change of state. The heat energy required to change the state of a substance is known as latent heat.

We can normally define two latent heats for a substance: one known as the latent heat of fusion, which refers to the change of a substance from the solid to the liquid state and vice versa; and the second known as the latent heat of vaporisation, which refers to the change from the liquid to the gaseous state. In both cases, no change in temperature is observed when the substance is changing its state.

The term fusion is used to describe the melting of a substance which occurs at a specific temperature known as the melting or freezing point temperature (for example the freezing point of water is also the melting point of ice). When a substance is melted, heat must be supplied. When a substance freezes, heat is given out. The term vaporisation is used to describe the change of a substance from the solid or the liquid state to a gas or vapour. A vapour is a gas, but 'vapour' is normally used to describe specifically a gas of a substance which exists mainly as a solid or liquid at normal temperatures and atmospheric pressure, e.g. water vapour, petrol vapour, alcohol vapour, scent. Some substances, e.g. naphthalene, change directly from the solid to the vapour state and in these cases, where the liquid phase is absent, we describe the vaporisation as sublimation. When a substance is vaporised, heat must be supplied to effect the transition. When a vapour condenses back to liquid or solid, heat is evolved. During both transitions there is normally no change in temperature—hence, as in the case of fusion, latent heat is either absorbed or evolved without any change in temperature.

(b) Temperature–time graphs for substances changing state

The distinction between sensible heat and latent heat is shown graphically in fig. 8.2. In (a) we have plotted a graph of temperature versus time for the transition of ice to water, and water to steam, for the case when heat is supplied at a uniform rate. Suppose we have, initially, a quantity of ice at −20 °C. Over the range AB, heat is supplied as sensible heat, the sensible heat raising the temperature from −20 °C to 0 °C.

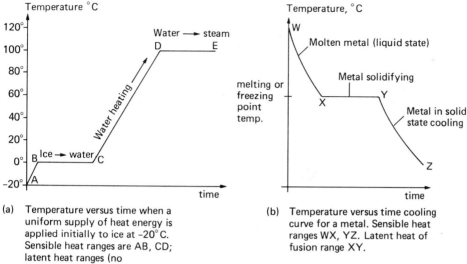

(a) Temperature versus time when a uniform supply of heat energy is applied initially to ice at –20°C. Sensible heat ranges are AB, CD; latent heat ranges (no temperature change) are BC, DE.

(b) Temperature versus time cooling curve for a metal. Sensible heat ranges WX, YZ. Latent heat of fusion range XY.

Fig. 8.2

0 °C is the melting point of ice and heat energy is now required to melt the ice to water. Thus, over the range BC, the heat supplied is known as latent heat. Temperature remains constant at 0 °C until all the ice is melted. Subsequently, over the range CD, sensible heat is absorbed and the water is raised from 0 °C to 100 °C, the boiling point of water. Latent heat (greatly in excess of the value required to melt the ice) must now be supplied to convert the water to steam. No change of temperature from 100 °C will occur until all the water has been vaporised to steam. Thus, we obtain the flat range DE. Finally, although not shown on the graph, the steam absorbs sensible heat and its temperature rises. In this final range, we say we have superheated steam.

Figure 8.2(b) shows the cooling graph for a pure metal initially in its liquid or molten state and above its melting point. Sensible heat is given out over the range WX; latent heat is given out over the range XY as the metal solidifies, and the temperature remains constant; finally, more sensible heat is given out over the range YZ as the solid metal cools.

8.4
Most materials expand or contract with temperature change: the effect of temperature changes on the physical dimensions of solids, liquids and gases

Most substances, whether in the solid, liquid, or gaseous state, expand when heated, that is as their temperature rises, and contract on being cooled. One notable exception to this rule is water, which actually contracts in volume as it is heated from 0 °C to 4 °C. However, over the range 4 °C to 100 °C, it expands with increasing temperature. Expansion and contraction occur in all three dimensions, but for solid materials we may quantify the effect of expansion in three ways:

(1) Linear expansion, which refers to the changes in length with temperature, and is of practical importance when dealing with wires, rods, cables, rails, etc.

(2) Area expansion, which refers to changes in area with temperature, and is useful when dealing with thin sheets of materials.

(3) Volume expansion which refers to changes in volume with temperature.

Since liquids and gases cannot 'hold' their shape, we can only really quantify expansion effects as a change in volume with temperature, although in the case of a gas, we must also consider pressure.

It is found, experimentally, that the length of many solid materials varies with temperature over a given temperature range according to the formula:

$$l_2 = l_1[1 + \alpha(T_2 - T_1)]$$

where l_2 – length at T_2 °C, l_1 – length at T_1 °C, and α (Greek letter alpha) is a constant for a

Table 8.2　Coefficients of linear and volume expansion of some substances at $20\,°C$

Substance	Coefficient of linear expansion, α (K^{-1} or $°C^{-1}$)	Coefficient of volume expansion, γ (K^{-1} or $°C^{-1}$)
Aluminium	24×10^{-6}	72×10^{-6}
Brass (60% Cu, 40% Zn)	20×10^{-6}	60×10^{-6}
Copper	17×10^{-6}	51×10^{-6}
Steel	12×10^{-6}	36×10^{-6}
Glass	$7\text{–}10 \times 10^{-6}$	$20\text{–}30 \times 10^{-6}$
Wood (along grain)	$3\text{–}5 \times 10^{-6}$	$70\text{–}130 \times 10^{-6}$
(across grain)	$35\text{–}60 \times 10^{-6}$	
Invar-steel alloy	2×10^{-6}	6×10^{-6}

given material. α is known as the coefficient of linear expansion. Using the formula we have:

$$l_2 = l_1 + \alpha l_1 (T_2 - T_1)$$

so change in length $l_2 - l_1 = \alpha l_1 (T_2 - T_1) = \alpha \times$ (length at $T_1\,°C$) × (temperature change) and

$$\alpha = \frac{l_2 - l_1}{l_1(T_2 - T_1)},$$

the units of α being per degree $°C$ or kelvin, i.e. $°C^{-1}$ or K^{-1}. It is usual practice to take l_1 as the reference length at $T_1 = 20\,°C$, although $0\,°C$ is also used.

The volume expansion of solids, liquids, and gases (at constant pressure) is given by a very similar formula:

$$v_2 = v_1[1 + \gamma(T_2 - T_1)]$$

where v_2 = volume at T_2, v_1 = volume at T_1 and γ (Greek letter gamma) is a constant for a given substance. γ is known as the coefficient of volume expansion.

The expansion constants vary slightly with temperature and it is usual practice to quote α and γ as the average value over a given temperature range. Table 8.2 gives values of α and γ for some common substances at $20\,°C$.

For gases we must be very careful and take into account changes in pressure when dealing with expansion or contraction effects. If a gas is maintained at a constant pressure, for example atmospheric pressure, then the above formula applies. In fact if we quote temperature in kelvin, i.e. $T\,K = 273 + T\,°C$, we have the very simple relationship of the volume of a gas being directly proportional to absolute temperature, provided that the pressure remains constant, so

$$v_1 \propto T_1(K), \quad v_2 \propto T_2(K)$$

hence

$$\frac{v_1}{v_2} = \frac{T_1(K)}{T_2(K)} \quad \text{or} \quad v_2 = \frac{T_2(K)}{T_1(K)}v_1 = \frac{273 + T_2\,°C}{273 + T_1\,°C}v_1$$

Example 8.4
Steel rails of length 10 m are laid end to end with a small gap to allow for expansion and contraction. If the maximum range in temperature variation is from $-10\,°C$ to $+40\,°C$, calculate the minimum gap that should be left. The coefficient of linear expansion of steel is $\alpha = 12 \times 10^{-6}\,°C^{-1}$.

Solution
Length at $40\,°C$,

$$l_{40} = l_{20}[1 + \alpha(40 - 20)] = l_{20}(1 + 20\alpha)$$

length at $-10\,°C$,

$$l_{-10} = l_{20}[1 + \alpha(-10 - 20)] = l_{20}(1 - 30\alpha)$$

and therefore total change in length from $+40\,°C$ to $-10\,°C$ is

$$
\begin{aligned}
l_{40} - l_{-10} &= l_{20}(1 + 20\alpha) - l_{20}(1 - 30\alpha) = 50\alpha l_{20} \\
&= 50 \times 12 \times 10^{-6} \times 10 \\
&= 6000 \times 10^{-6}\,m = 6\,mm
\end{aligned}
$$

Thus the minimum gap should be at least 6 mm. *Note:* Suppose no gap were left and an expansion of 1 mm was prohibited; in fact this corresponds to less than a $10\,°C$ rise in temperature. Then the compressive forces experienced in the rail cross-section would be very great, i.e.

$$\text{force} = A \times E \times \text{strain}$$

where

A = cross-sectional area = $100\,\text{cm}^2$
$\quad = 10^{-2}\,\text{m}^2$ (say)

E = Young's modulus for steel
$\quad = 212 \times 10^9\,\text{N/m}^2$

$$\text{strain} = \frac{x}{l} = \frac{1\,\text{mm}}{10\,\text{m}} = 10^{-4}$$

so

$$\text{force} = 10^{-2} \times 212 \times 10^9 \times 10^{-4} = 212 \times 10^3\,\text{N}$$

Problems 8: Heat

1.   Distinguish between heat and temperature and state the SI units of these quantities. Convert (a) $150\,°\text{C}$, $-39\,°\text{C}$, $1000\,°\text{C}$ to kelvin; (b) $50\,\text{K}$, $273\,\text{K}$, $1500\,\text{K}$ to degrees celsius.

2.   Define the specific heat capacity of a substance.

Calculate the amount of heat required to raise the temperature of $400\,\text{kg}$ of water in a copper tank of mass $50\,\text{kg}$ from $5\,°\text{C}$ to $40\,°\text{C}$. Neglect any heat losses to the surroundings. The specific heat capacities of water and copper are respectively $4180\,\text{J}\,\text{kg}^{-1}\,°\text{C}^{-1}$ and $385\,\text{J}\,\text{kg}^{-1}\,°\text{C}^{-1}$.

3.   A $0.25\,\text{kg}$ mass of iron at an unknown temperature is immersed in $1.2\,\text{kg}$ of water at an initial temperature of $18\,°\text{C}$. It is observed that the temperature of the water rises to $29\,°\text{C}$. Neglecting any heat losses to the surroundings, calculate the initial temperature of the mass of iron. The specific heat capacities of water and iron respectively are $4186\,\text{J}\,\text{kg}^{-1}\,°\text{C}^{-1}$ and $460\,\text{J}\,\text{kg}^{-1}\,°\text{C}^{-1}$.

4.   Distinguish between sensible and latent heat. Sketch the form of the temperature–time graph showing the cooling curve of a metal initially in a molten state.

5.   Sketch and explain the temperature–time graph for water changing from ice to water, water to steam.

Determine how long it will take for $2\,\text{kg}$ of water initially at $10\,°\text{C}$ to boil in a $2\,\text{kW}$ kettle. Neglect heat losses and the energy required to heat the metal of the kettle. The specific heat capacity of water is $4186\,\text{J}\,\text{kg}^{-1}\,°\text{C}^{-1}$.

6.   The length of a steel beam at a temperature $T\,°\text{C}$ is given by

$$l = 5[1 + 12 \times 10^{-6}(T - 20)].$$

Calculate its length at $-20\,°\text{C}$, $0\,°\text{C}$, and $+40\,°\text{C}$.

It is known that the compressive force set up in the beam is given by

$$F = 40 \times 10^6 x \text{ newtons}$$

when the beam is compressed by x metres. Calculate the compressive force set up in the beam when it is clamped at each end at $20\,°\text{C}$ and its temperature is then raised to $100\,°\text{C}$.

C Electricity
9 Resistive electrical circuits

The expected learning outcome of this chapter is that the student solves problems related to current, potential difference and resistance for simple series and parallel resistive circuits.

Introduction: The selection and use of preferred unit prefixes in accordance with SI

In the study of electricity we are concerned with two basic quantities: electrical current, which will be denoted by the symbol I, and voltage, which will be denoted by the symbol V. We shall also be meeting components which exhibit the property of restricting current and converting electrical energy to heat. This property is known as resistance, and we shall denote resistance by the symbol R. We shall be explaining the meaning of these quantities, defining their units, and using I, V, and R in electrical problems. However, first let us explain the way we write down the numerical value of these quantities (and, of course, other electrical and non-electrical quantities), by employing decimal prefixes.

We use prefixes to denote decimal multiples (factors of 10) or sub-multiples (fractions of 10) of units of current, voltage, resistance, etc. This enables us to write down the value of physical quantities in a concise and clear manner. The symbols, names, and meanings of the SI preferred prefixes are listed in Table 9.1.

Let us now illustrate how these prefixes are used, by the following examples: I is current and has the units of amperes, the abbreviated symbol for amperes being A; V is voltage and has the units of volts, symbol V; R is resistance and has the units of ohms, and is abbreviated by the Greek letter Ω, called omega.

(1) $I = \dfrac{7}{1000} = 0.007 = 7 \times 10^{-3}\,\text{A}$

$\quad = 7\,\text{mA}$

m is the $\times 1/1000$ or $\times 10^{-3}$ prefix, A the units of current, and mA is the abbreviation for milliamperes.

(2) $V = 1400\,\text{V}$ may be written as
$\quad V = 1.4 \times 1000$ or $1.4 \times 10^3\,\text{V}$
$\quad = 1.4\,\text{kV}$

k is the $\times 1000$ or $\times 10^3$ prefix, V the unit of voltage, and kV the abbreviation for kilovolts.

(3) $R = 3\,000\,000 = 3 \times 10^6\,\Omega$
$\quad = 3\,\text{M}\Omega$

M is the times a million prefix, Ω the unit abbreviation for resistance, and MΩ the abbreviation for megaohms.

(4) $15\,\mu\text{A} = 15 \times 10^{-6}\text{A}$ (fifteen microamperes)
$\quad 10\,\text{ns} = 10 \times 10^{-9}\text{s}$ (ten nanoseconds)
$\quad 3\,\text{GN} = 3 \times 10^{9}\text{N}$ (three giganewtons)
$\quad 100\,\text{pW} = 100 \times 10^{-12}\text{W}$ (hundred picowatts)

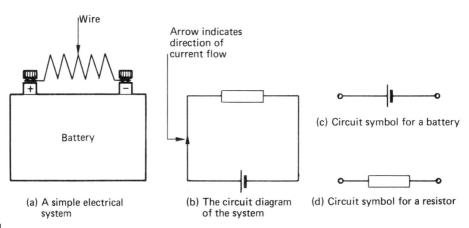

(a) A simple electrical system

(b) The circuit diagram of the system

(c) Circuit symbol for a battery

(d) Circuit symbol for a resistor

Fig. 9.1

Table 9.1   Names and symbols for the preferred SI unit prefixes

Symbol	Name	Meaning (i.e. factor by which unit is multiplied)
T	tera	$\times 10^{12}$ (a million million times)
G	giga	$\times 10^{9}$ (a thousand million times)
M	mega	$\times 10^{6}$ (a million times)
k	kilo	$\times 10^{3}$ (a thousand times)
m	milli	$\times 10^{-3}$ (a thousandth)
μ	micro	$\times 10^{-6}$ (a millionth)
n	nano	$\times 10^{-9}$ (a thousand millionth)
p	pico	$\times 10^{-12}$ (a million millionth)

9.1

Standard symbols for electrical components when drawing circuit diagrams

Figure 9.1(a) shows a sketch of a simple electrical system consisting of a battery whose terminals are connected to the ends of a length of wire. The battery acts as a source of energy generating current and the wire acts as a conductor for this current. To simplify having to draw a physical picture of the components making up an electrical system (which would be very arduous, anyway, except in the simplest of cases), we use universally recognised symbols to denote the

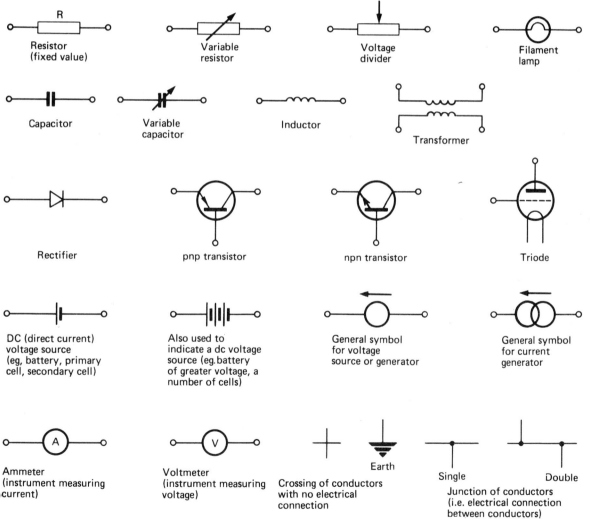

Fig. 9.2

components and draw the system in the form of a diagram—known as a circuit diagram.

The circuit diagram representing fig. 9.1(a) is drawn in fig. 9.1(b). The battery is represented by the symbol shown in fig. 9.1(c) and the wire which acts as a resistance impeding the flow of current, by the symbol shown in fig. 9.1(d). A fuller list of symbols used for electrical components in circuit diagrams is given in fig. 9.2.

The main symbols that we shall be using in this book will be that for the battery (source of electrical energy) and that for the resistor (an electrical component exhibiting the property of resistance). The symbols for the other electrical components included in fig. 9.2 will be used in the later stages of your study of electricity.

9.2
Some basic concepts concerning electrical current

(a) *Electrical current and charge and the units of current and charge*

The specific learning objective of this section is that you should understand the meaning of electrical current, know its units and appreciate that continuous current can only flow if the circuit is complete.

However, it is useful to begin with a short discussion on the origin of electrical charge and current. All matter is composed of basic building blocks known as atoms. Going one stage further in atomic theory concepts, we can picture an atom as composed of a positively charged central core, known as the nucleus, surrounded by tiny negatively charged particles rotating around the nucleus in a series of fixed orbits (in an analogous way to the planets rotating around the sun). These negatively charged particles are known as electrons and each carries the smallest known amount of electrical charge. In SI the unit of electrical charge is the coulomb, abbreviation C. The charge of an electron has a magnitude of 1.6×10^{-19} C.

In an electrically neutral atom the total positive charge of the nucleus is equal to the total negative charge of the electrons. Thus, if electrons are removed from a substance, the substance becomes positively charged, whilst if there is an excess of electrons in a substance, it becomes negatively charged. Atoms, or groups of atoms, with a net positive or a net negative charge are known as ions.

The electrons orbiting closest to the nucleus are bound tightly to the atom by attraction forces due to the positively charged nucleus. In an insulator, all the orbital electrons are held securely, but in a conductor, the outermost electrons are bound very weakly and can break free from the atom and drift in a random manner through the conductor. These electrons are called free electrons.

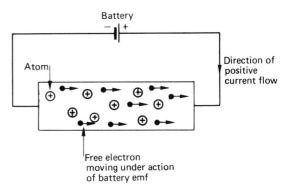

Fig. 9.3

If an electrical force is set up in a conductor, which may be achieved, for example, by connecting a battery across the conductor ends (as shown in fig. 9.3), the free electrons are acted upon by electrical forces and a uniform movement is superimposed on the previous random paths of the free electrons. This flow of electrons due to the applied source (the battery) constitutes an electrical current. In general, most current flow is due to free electrons, but in gases and liquids, current can be constituted by the flow of ions (e.g. atoms which have either gained or lost one or more free electrons). The direction in which current flows in a circuit is conventionally regarded as that in which positive charge will move, that is, it is opposite in direction to the flow of electrons.

Finally let us summarise and state that electrical current is the rate of flow of electrical charge, i.e. the number of coulombs of charge passing a given position per second. Current is measured in the SI units of amperes, symbol A. One ampere is equal to the flow of one coulomb per second. In terms of the flow of free electrons we have

$$\begin{aligned}
1 \text{ ampere} &= 1 \text{ coulomb per second} \\
&= 1 \div (\text{electron charge}) \\
&= 1 \div (1.602 \times 10^{-19}) \\
&= 6.24 \times 10^{18} \text{ electrons per second.}
\end{aligned}$$

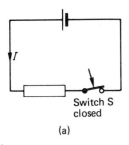

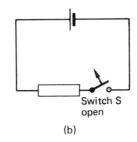

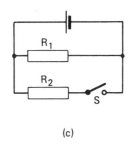

Fig. 9.4

Common sub-multiple units of current are the milliampere (mA) and the microampere (μA):

$$1\,mA = 0.001 \text{ or } 10^{-3}\,A, \quad 1\,\mu A = 10^{-6}\,A$$

(b) For a continuous current flow a complete circuit is necessary

For a continuous flow of current a complete circuit is necessary. This means there must be at least one unbroken conduction path when tracing a path through the circuit from one terminal of the battery (or other energy source) to the other. Any break in the conducting wires joining up the components in a series circuit, for example, prevents a flow of electrical current. Figure 9.4(a) shows a circuit in which the current is continuous, when the switch S is closed. Figure 9.4(b) shows a circuit where a break is made by opening the switch S. The open switch breaks the conduction path for current (current does not normally flow through air, since air acts as an excellent insulator) and so no current flows. In the circuit of fig. 9.4(c) current will flow through the component R_1 since a continuous conduction path exists to and from the battery via R_1. However, no current flows through the component R_2 until switch S is closed.

9.3
Current flows due to the existence of a potential difference (voltage) between two points in an electrical conductor

We have seen that a continuous circuit is necessary for an electrical current to flow, and that electrical current is the transportation or flow of electrical charge. However, no current will flow through a conductor unless a source of energy is supplied to create a force in the conductor, and therefore cause a resultant flow of charges in one direction. If no such source is present, the resultant motion of the large number of free electrons in a conductor will be as much one way as the other, with the net effect that no current flows.

The source of electrical energy we have so far used in our circuit diagrams is the battery. Batteries (e.g. dry batteries, accumulators, or car batteries) produce electrical energy by the conversion of chemical energy. Some examples of other sources of electrical energy are electro-mechanical machines which produce electrical energy from mechanical energy of rotation; and solar cells which convert light energy directly into electrical energy.

As a measure of the ability of a source to effect the flow of current in a circuit, we use the term of electromotive force, usually abbreviated to e.m.f. If the e.m.f. remains constant (i.e. does not vary its strength with time), it will produce a steady current flow. Steady current is known as direct current, and has the abbreviation d.c. However, if the e.m.f. of the source varies with time in its magnitude (strength) and its direction of action, this will cause a varying current to flow in a circuit which will vary continuously in strength and also change its direction of current flow. In our homes the electrical supply is of this type and produces alternating current. The abbreviation of alternating current is a.c. Graphs of a.c. and d.c. currents versus time are shown in fig. 9.5.

We now consider the concept of potential difference and the fact that a potential difference is necessary to cause current to flow in a conductor. To explain this we shall use the simple circuit of fig. 9.6 where a d.c. source of e.m.f. is connected across the terminals XY of a conductor. Energy is continually supplied from this source to maintain current flow through the conductor. The e.m.f. of the source is defined by the energy supplied by the source in transporting one unit of positive charge (1 coulomb) from the positive side of the source (often called the

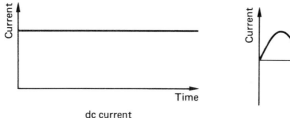

Fig. 9.5

positive polarity terminal and represented on the circuit diagram by $+$) through the conductor to the negative polarity terminal ($-$). If we consider any two points A and B in the conductor, the work done in transporting the unit charge from A to B is called the potential difference or voltage between A and B. The work done, of course, is supplied by the source. For current to flow between A and B there must be a potential difference between A and B in the same way as there must be a height difference between two points in a drainage duct or a river if a flow of water is to be maintained between these two points. In the latter case, gravitational forces do work in transporting water. In the electrical case, our electrical source supplies the energy to transport the electrical charge.

In circuit diagrams, the direction of current (i.e. the direction in which positive charge flows) is indicated by an arrow, this arrow or arrowhead normally being marked in on one of the connecting wires of the circuit, as shown, for example, in fig. 9.6. Current always flows from a higher voltage point to a lower voltage point. The potential difference or voltage between two points is also indicated by an arrow, as shown in fig. 9.6, the higher voltage point being indicated by the arrow tip. Thus, in fig. 9.6 the arrow shows that A is at higher voltage than B.

The unit of potential difference (voltage)

The SI unit of potential difference or voltage is

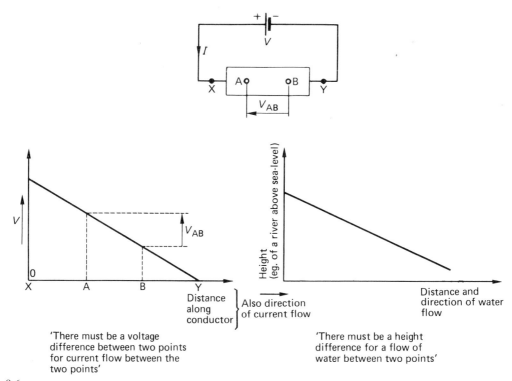

Fig. 9.6

the volt, which has the abbreviation V. Potential difference and voltage have identical meaning. The potential difference between two points is the work done in taking 1 coulomb of charge from one point to the other. Thus, remembering that the SI unit of work and energy is the joule (J), we have

$$1 \text{ volt} = \frac{1 \text{ joule}}{1 \text{ coulomb}}$$

The e.m.f. of an electrical energy source is also measured in volts. Multiple and sub-multiple units of the volt which are fairly frequently used are the kilovolt (kV), the millivolt (mV), and the microvolt (μV), e.g.

$$10 \text{ kV} = 10\,000 \text{ V},$$
$$50 \text{ mV} = 0.050 \text{ V},$$
$$25 \text{ } \mu\text{V} = 0.000\,025 \text{ V}$$

9.4
The measurement of current using an ammeter and potential difference using a voltmeter

An instrument that measures current is known as an ammeter. The ammeter has two terminals one normally marked with a +, and one marked with a −, and a scale calibrated in amperes or sub-multiple units of amperes such as milliamperes (mA) or microamperes (μA). Sketches of some common forms of ammeters (and voltmeters), together with their circuit symbols, are shown in fig. 9.7. To measure the current in a circuit, we must break the circuit at a suitable point and connect the + terminal of the ammeter to the more positive side (high voltage side) of the circuit and the − terminal to the other side. Examples of the connection of an ammeter to measure current in the various components making up the circuit of fig. 9.8(a) are shown in (b), (c), and (d). Note that the ammeter is always connected in series with the component.

An instrument that measures voltage is known as a voltmeter. It is of almost identical construction to the ammeter; in fact it only requires a relatively simple modification to convert an ammeter to a voltmeter and vice versa. It has two terminals, one +, one −, at which external connections are made and a scale calibrated in volts or multiples or sub-multiples of volts, such as kilovolts (kV) or millivolts (mV). To measure the voltage across a circuit component, we

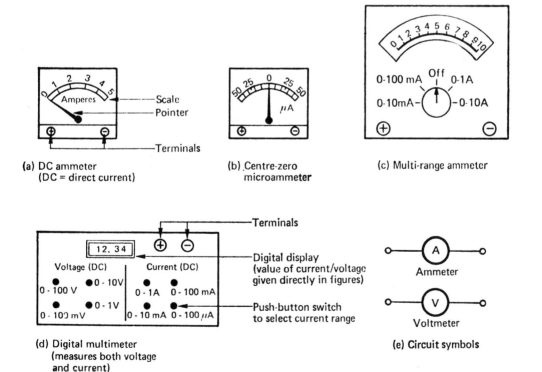

(a) DC ammeter
(DC = direct current)

(b) Centre-zero microammeter

(c) Multi-range ammeter

(d) Digital multimeter
(measures both voltage and current)

(e) Circuit symbols

Fig. 9.7

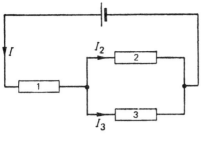

(a) Circuit diagram

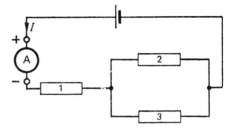

(b) Connection of ammeter to
measure total circuit current I

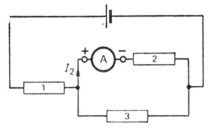

(c) Connection of ammeter to
measure current I_2 in component 2

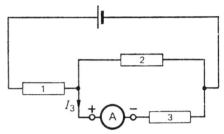

(d) Connection of ammeter to
measure current I_3 in component 3

Fig. 9.8

connect the voltmeter across the component (i.e. in parallel with the component) with the + terminal at the higher voltage side and the − terminal connected to the lower voltage side. Examples of the connection of a voltmeter to measure the voltage across the various components of fig. 9.9(a) are shown in (b) and (c).

9.5
The drawing of graphs of the relationship between voltage and current using experimental data for (a) a linear resistor, and (b) a non-linear resistor

(a) *Current versus voltage graph for a linear resistor*

A graph of the current I through a resistor versus voltage V across it may be obtained experimentally using the circuit shown in fig. 9.10(a). I is measured by an ammeter and V by a voltmeter. Note that the ammeter is connected in series with the resistor and the voltmeter is connected across the terminals of the resistor. The voltage across the resistor can be varied in suitable steps by either changing the voltage of the source, as shown in fig. 9.10(b), or by using a variable resistor in series with the given resistor and battery as shown in fig. 9.10(c).

In such an experiment, the following values were obtained (opposite). Table 9.2(a) refers to the values of voltage and current obtained using the circuit of fig. 9.10(a). Table 9.2(b) refers to

Table 9.2

V volts voltage across resistor R	I milliamperes current through resistor R	V volts	I milliamperes
0	0	0	0
1	40	−1	−40
2	80	−2	−80
3	120	−3	−120
4	160	−4	−160
5	200	−5	−200
(a)		(b)	

the results obtained when we reverse the polarity of the source to drive current through the resistor in the opposite direction. Remember, when we reverse the battery terminals, we must also reverse the connections of both ammeter and voltmeter as shown in fig. 9.10(d).

The graph of I versus V is plotted in fig. 9.11

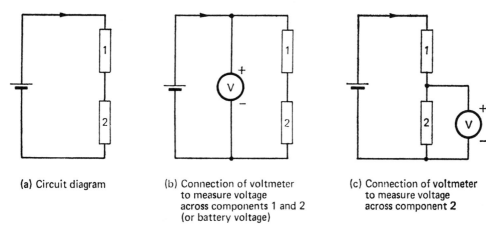

(a) Circuit diagram

(b) Connection of voltmeter to measure voltage across components 1 and 2 (or battery voltage)

(c) Connection of voltmeter to measure voltage across component 2

Fig. 9.9

with the y or vertical axis chosen as the current axis and the x or horizontal axis as the voltage axis. In plotting this graph we chose the following scales: 40 mA per division for the current and 1 V per division for the voltage. Each point should be plotted in turn; for example,

when $V = 4\,V$ we move 4 divisions along the horizontal axis, and as the current corresponds to 160 mA, we move 4 divisions vertically. We then mark a X (cross) at this point. After all points are plotted, a smooth curve should be drawn to connect up all points of our graph. In

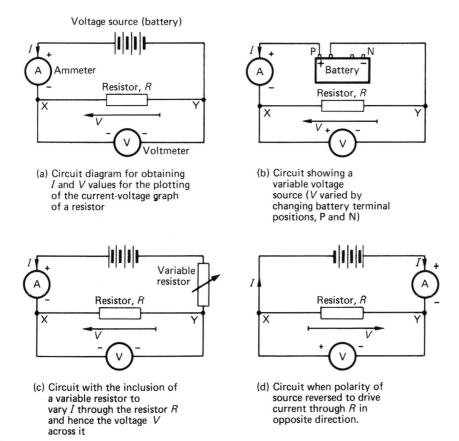

(a) Circuit diagram for obtaining I and V values for the plotting of the current-voltage graph of a resistor

(b) Circuit showing a variable voltage source (V varied by changing battery terminal positions, P and N)

(c) Circuit with the inclusion of a variable resistor to vary I through the resistor R and hence the voltage V across it

(d) Circuit when polarity of source reversed to drive current through R in opposite direction.

Fig. 9.10

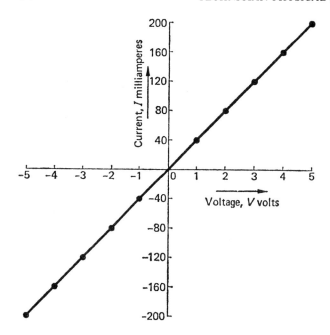

Fig. 9.11   The graph of current I versus voltage V for the linear resistor R whose V–I data are given in Table 9.2

this case our task is easy since the graph is a straight line.

There are some very important observations to be made from our graph:

(1) The I versus V graph is a straight line.
(2) The I versus V graph passes through the origin, i.e. when $V = 0$, $I = 0$.
(3) When we reverse the voltage across the resistor, we obtain identical values of current, although this flows in the opposite direction. On the graph these readings are plotted as negative values.

A resistor whose I versus V graph is a straight line passing through the origin, is known as a linear resistor. Most metal alloy and carbon resistor materials act as linear resistors and these components are extensively used in electrical and electronic circuits.

(*b*) *A current versus voltage graph for a non-linear resistor, such as a lamp*

There are many electrically conducting materials which do not act as linear resistors, i.e. when we plot their I versus V graph, we do not obtain a straight line.

Let us investigate the I versus V graph for a non-linear resistor choosing a simple torch lamp bulb as an example. A simple circuit which may be used to obtain current and voltage values for the lamp bulb is shown in fig. 9.12. The results obtained are listed in Table 9.3.

The graph of I versus V for the lamp bulb is plotted in fig. 9.13. The scale for current is 100 mA per division and the scale for voltage is 2 V per division. Each point is plotted in turn and the smooth graph drawn through all the points. We observe straight away that the graph is not a straight line and hence we can state that the lamp bulb acts as a non-linear resistor element.

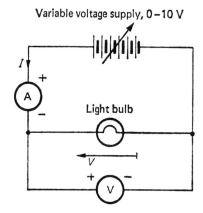

Fig. 9.12   Circuit diagram for obtaining I and V values for the plotting of the current versus voltage graph of a light bulb (non-linear) resistor

Example 9.1
The following results were obtained for the

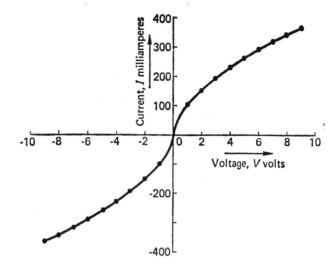

Fig. 9.13 The graph of current I versus voltage V for the lamp bulb, whose V–I data are given in Table 9.3

voltage V across and the current I flowing through two resistance components.

Component A

V (V)				-5		-4		-3		-2
I (mA)				-100		-80		-60		-40

V (V)	-1	0	$+1$	2	3	4	5
I (mA)	-20	0	20	40	60	80	100

Component B

V (V)			-1	-0.5	0	$+0.5$
I (mA)				0	5 0	0.08

V (V)	0.6	0.7	0.8	0.9	1	1.1
I (mA)	1	3	7.5	16	26	40

Draw the graphs of the relationship of current versus voltage for component A and B and hence state which component is a linear and which is a non-linear resistance element.

Determine (a) the current through component A when the voltage across A is equal to 3.5 V, (b) the voltage across component B when the current flowing through it is 20 mA.

Solution

The graph of I versus V for component A is drawn in fig. 9.14(a), and the graph for component B is drawn in fig. 9.14(b). The relationship between I and V for component A is a straight line, hence component A is a linear

Table 9.3

V volts voltage across lamp bulb	I milliamperes current through lamp bulb	V volts	I milliamperes
0	0	0	0
0.5	72	-0.5	-72
1	102	-1	-103
2	152	-2	-152
3	191	-3	-191
4	233	-4	-232
5	264	-5	-264
6	292	-6	-292
7	319	-7	-318
8	343	-8	-343
9	366	-9	-366

element. Component B is a non-linear element since fig. 9.14(b) shows that the plot of I versus V is not a straight line.

(a) We may find the current through A when $V = 3.5$ V by using the I v. V graph. First locate 3.5 V on the voltage axis and draw a vertical line as shown in fig. 9.14(a). This line cuts the I–V straight line of component A at the point Q (also shown). Draw a horizontal line through Q and read off value of current where this line cuts I axis. From fig. 9.14(a), this gives

$I = 70$ mA ... the current through A, when $V = 3.5$ V

(b) On the I v. V curve of fig. 9.14(b) draw horizontal line through $I = 20$ mA. This line cuts

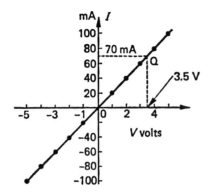

(a) *I* v *V* graph for component A

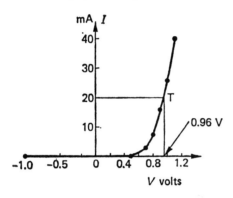

(b) *I* v *V* graph for component B

Fig. 9.14

curve at T. Draw vertical line through T and read off value of V where it cuts the voltage axis. This gives (as shown in fig. 9.14(b))

$$V = 0.96\,\text{V (approximately), when } I = 20\,\text{mA}$$

9.6
The definition of resistance as the ratio of potential difference (voltage) across a resistor to the current through it, and the description of resistance as that property of the conductor that limits current

We have used the terms resistance and resistor in the previous sub-section and shall now consider their formal definitions. The British Standard definition of resistance is as follows:

(1) Resistance is that property of a substance which restricts the flow of electricity through it associated with the conversion of electrical energy to heat.

(2) The magnitude of this restriction.

The term resistor is given to components which have the property of resistance in that they limit the flow of electrical current and dissipate electrical energy as heat. It is important to remember resistance as the property of restricting flow together with the conversion of electrical energy to heat, since there are other electrical components which can restrict current flow without necessarily converting electrical energy to heat, e.g. inductors and capacitors can restrict alternating current without converting any electrical energy to heat.

The British Standard definition of resistance is very general, but does not give a quantitative value to the resistance of a conductor. The value of the resistance of a conductor is defined as

$$R = \frac{V}{I}$$

where R is the symbol used to denote resistance,

V is the voltage across the ends of the conductor, units volts (V),

I is the current flowing through the conductor, units amperes (A).

Resistance is measured in units called ohms, denoted by the Greek letter omega, Ω. Note that

$$1\,\text{ohm} = \frac{1\,\text{volt}}{1\,\text{ampere}}, \quad \text{i.e. } \Omega = \text{V/A}$$

Some examples of resistor components are shown in fig. 9.15. Figure 9.15(a) is an example of a fixed resistor of the type commonly used in electronic circuits. It consists basically of a composition of suitable resistive material (e.g. moulded carbon) with connecting leads firmly attached to its ends. Figure 9.15(b) shows a wire-wound fixed resistor, which consists of resistive wire (e.g. manganin or constantan wire) wound on an insulating former and normally covered with a layer of vitreous enamel to help with the conduction of heat away from the component. Figure 9.15(c) shows two examples of a variable resistor. They consist of bare (uninsulated) resistance wire wound on an insulating former between two fixed terminals A and C. B is a movable spring contact whose position can be varied manually. Thus the resistance between terminals A and B can be varied between zero and the full value of the complete resistor coil of wire when B is moved from A to C.

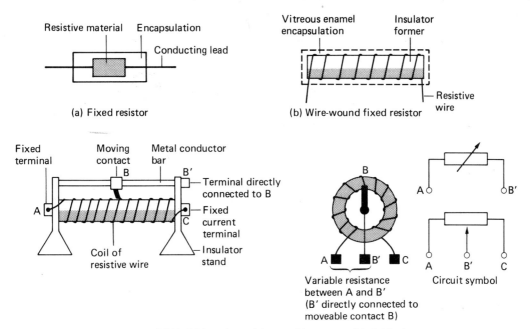

(a) Fixed resistor

(b) Wire-wound fixed resistor

Variable resistance between A and B'
(B' directly connected to moveable contact B)

Circuit symbol

(c) Variable resistors (also useable as potential dividers)

Fig. 9.15

Example 9.2
Determine the resistance of the resistors shown in fig. 9.16.

Solution
For resistor (a),

$$R = \frac{30\,V}{3\,A} = 10\,\Omega$$

For resistor (b),

$$R = \frac{10\,V}{5\,mA} = \frac{10\,V}{5 \times 10^{-3}\,A} = 2 \times 10^3 = 2\,k\Omega$$

For resistor (c),

$$R = \frac{1.5\,V}{10\,\mu A} = \frac{1.5\,V}{10 \times 10^{-6}\,A} = 1.5 \times 10^5$$

$$= 150 \times 10^3 = 150\,k\Omega$$

9.7
Statement of Ohm's Law in terms of proportionality of current to voltage

The relationship between the voltage across a conductor and the current flowing through it was first experimentally investigated by Ohm. Ohm observed that at constant temperature the current flowing through a conductor was proportional to the voltage across it. This relationship is known as Ohm's Law, which is certainly the most famous and one of the most widely applied 'laws' in electricity.

Ohm's Law states:

The current I flowing in a conductor is directly proportional to the voltage V across its ends, provided physical conditions, such as temperature, remain constant.

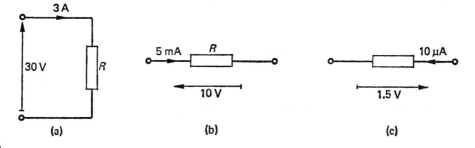

(a) (b) (c)

Fig. 9.16

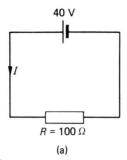
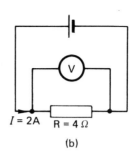
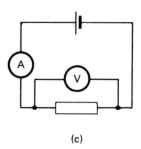

(a)   (b)   (c)

Fig. 9.17

The law may be expressed by the formula,

$$I = \frac{V}{R} \quad \text{or} \quad V = RI$$

where the constant of proportionality R is the resistance of the conductor.

Although this law is of utmost importance, it should be noted that it is not universal in its application. It is a condition closely obeyed by most metallic conductors, some non-metallic conductors, and some salt, and other electrolytic, solutions. However, some good conductors of electricity and semi-conductor materials (which are used to make diodes and transistors) do not obey Ohm's Law. For example, we saw in Section 9.5 that although the current in the lamp bulb increased as the voltage across it increased, the current through the lamp was not directly proportional to the voltage across it. Conductors which do not obey Ohm's Law are termed non-linear.

9.8
The solution of simple problems using Ohm's Law

Let us see how we may use Ohm's Law to solve problems of current flow in linear resistors. If we are given the value of resistance R and the voltage across the resistor V, then the current through the resistor is

$$I = \frac{V}{R}$$

Alternatively, if we know R and the current I, the voltage across the resistor is

$$V = RI$$

Finally, if we know V and I for the resistor, then

$$R = \frac{V}{I}$$

Thus, to summarise, if we know any two of the values of R, V, and I, we can find the unknown quantity:

$$R, V \text{ given,} \quad \text{then} \quad I = \frac{V}{R}$$

$$R, I \text{ given,} \quad \text{then} \quad V = RI$$

$$I, V \text{ given,} \quad \text{then} \quad R = \frac{V}{I}$$

Example 9.3
Calculate:

(a) The current I in the circuit of fig. 9.17(a).
(b) The reading of the voltmeter connected across the $4\,\Omega$ resistor in fig. 9.17(b).
(c) The value of the resistor in the circuit of fig. 9.17(c), given that the voltmeter reads 2 V and the ammeter reads 10 mA.

Solution
(a) The voltage across the resistor $V = 40$ V. The value of the resistor $R = 100\,\Omega$. The current $I = 40/100 = 0.4$ A.
(b) Since $I = 2$ A and $R = 4\,\Omega$, the voltage across the resistor is

$$V = RI = 4 \times 2 = 8 \text{ V}$$

This is equal to the voltage read by the voltmeter, since the voltmeter is connected across R.
(c) The current I through the resistor = current measured by the ammeter, i.e.

$$I = 10 \text{ mA} = 10 \times 10^{-3} \text{ A}$$

The voltage V across the resistor = voltage measured by the voltmeter, i.e.

$$V = 2 \text{ V}$$

Thus the resistance R of the resistor is given by

$$R = \frac{V}{I} = \frac{2}{10 \times 10^{-3}} = \frac{2}{0.01} = 200\,\Omega$$

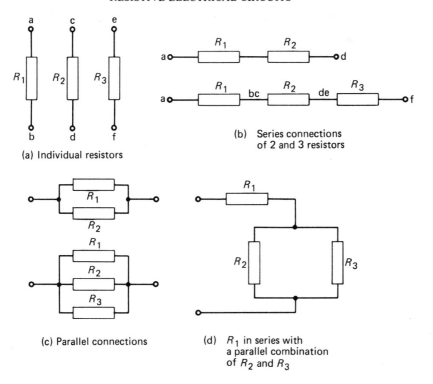

(a) Individual resistors

(b) Series connections of 2 and 3 resistors

(c) Parallel connections

(d) R_1 in series with a parallel combination of R_2 and R_3

Fig. 9.18

9.9
The difference between series and parallel connections of resistors

There are two basic ways of interconnecting resistors: either in series or in parallel. Figure 9.18 shows the difference between the series and parallel connection of resistors. In fig. 9.18(b) the resistors are joined in series, i.e. end-to-end connection. In fig. 9.18(c) the resistors are connected in parallel, i.e. the terminals are joined together at each end. In fig. 9.18(d) the resistor R_1 is connected in series with R_2 in parallel with R_3.

9.10
Properties of series circuits: (a) the current is the same in all parts of the circuit; (b) the sum of the voltages is equal to the total applied voltage

(a) The current is the same in all parts of a series circuit

In the series circuit of fig. 9.19(a) the battery is connected in series with the resistors R_1 and R_2 and these components will be almost invariably situated in air. Air is a non-conductor of electricity and thus current cannot escape from the circuit. The current I leaving the positive terminal of the battery enters R_1, leaves R_1, enters R_2, leaves R_2, and completes the circuit by entering the negative terminal of the battery, i.e. the current is the same in all parts of the circuit.

In general, we may state that the current is the same in all parts of a series circuit. For example, if in the series circuit of fig. 9.19(b) the current measured by ammeter is 2 A, then this is the value of current in all parts of the circuit with 2 A flowing through R_1, R_2, and R_3.

(b) The sum of the voltages in a series circuit is equal to the total applied voltage

In a series circuit, the sum of the individual voltages across the individual components is equal to the total applied voltage (i.e. the voltage impressed by the source across the circuit). Thus, in fig. 9.19(a) where the battery voltage V is the applied voltage and the voltage across R_1 is V_1 and across R_2 is V_2, we have $V = V_1 + V_2$.

In fig. 9.19(b) the applied voltage $V = 100$ V, the voltage across R_1 is $V_1 = 20$ V, across R_2 is

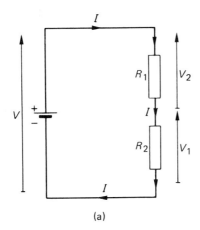

(a)

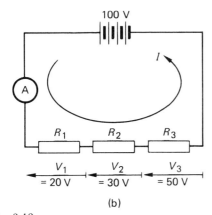

(b)

Fig. 9.19

$V_2 = 30$ V, and the voltage across R_3 is $V_3 = 50$ V,
so we have

$$V = V_1 + V_2 + V_3$$

i.e.

$$100 = 20 + 30 + 50$$

Example 9.4
The current in the series circuit of fig. 9.20 is $I = 5$ A. The applied source voltage is 40 V and the voltage across R_1 is 10 V. Calculate (a) the voltage V_2 across R_2, (b) the value of R_2.

Solution
(a) Since the applied voltage equals the sum of the voltage across R_1 and R_2, we have

$$40 = 10 + V_2$$

so

$$V_2 = 40 - 10 = 30 \text{ V}$$

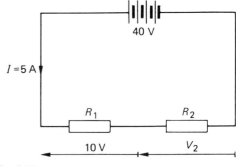

Fig. 9.20

(b) The current is the same in all parts of a series circuit, so the current $I = 5$ A passes through both R_1 and R_2. Thus, since the voltage across R_2 is $V_2 = 30$ V, we have

$$R_2 = \frac{V_2}{I} = \frac{30}{5} = 6 \,\Omega$$

9.11
The equivalent resistance of a number of resistors connected in series is given by the sum of the individual resistances

In the last two sub-sections, we learnt two very important facts about the properties of series circuits.

(1) The current is the same in all parts of the circuit.
(2) The sum of the individual voltages is equal to the applied voltage.

We shall now apply these properties to determine the total equivalent resistance R of the series circuit of fig. 9.21(a), i.e. the single resistor which when it replaces the series connection of resistors R_1, R_2, and R_3 in the circuit causes the identical current to flow. Thus, referring to fig. 9.21(b), the equivalent resistance of R_1, R_2, and R_3 in series is defined as

$$R = \frac{V}{I}$$

where V is the battery voltage across terminals A and B, and I is the current flow in the series circuit.
We shall show that

$$R = R_1 + R_2 + R_3$$

i.e. R equals the sum of the individual resistances.

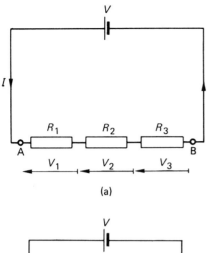

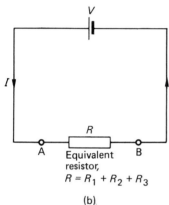

Fig. 9.21

Since the applied voltage V across A–B equals the sum of the individual voltages, we have

$$V = V_1 + V_2 + V_3$$

and on applying Ohm's Law to each of the individual resistors, we have

$$V_1 = R_1 I, \quad V_2 = R_2 I, \quad V_3 = R_3 I$$

Thus

$$V = V_1 + V_2 + V_3$$
$$= R_1 I + R_2 I + R_3 I = (R_1 + R_2 + R_3)I$$

and rearranging this equation to solve for I we obtain

$$I = \frac{V}{R_1 + R_2 + R_3} \equiv \frac{V}{R}$$

showing that

$$R = R_1 + R_2 + R_3$$

since the equivalent resistance R must give the same current when it replaces the 3-series resistors in the same circuit.

In general, the total resistance of any number of resistors connected in series is equal to the sum of their individual resistances.

9.12
The solution of problems for series circuits including the use of Ohm's Law

The following examples are concerned with series circuits and may be solved by applying Ohm's Law and remembering that in series circuits:

(a) the current is the same in all parts of the circuit,
(b) the total applied voltage is equal to the sum of the individual voltages,
(c) the total resistance is equal to the sum of the individual resistances.

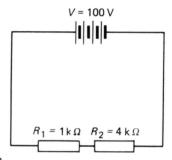

Fig. 9.22

Example 9.5
Calculate (a) the current I in the series circuit of fig. 9.22, (b) the ratio of the voltage V_1 across R_1 to the voltage V_2 across R_2.

Solution
(a) The total resistance of the circuit,

$$R = R_1 + R_2 = 1 + 4 = 5 \, k\Omega$$

Now the applied voltage,

$$V = 100 = RI$$

so the current,

$$I = \frac{V}{R} = \frac{100}{5} = 20 \, mA$$

(b) The voltage $V_1 = R_1 I$ and the voltage $V_2 = R_2 I$, so

$$\frac{V_1}{V_2} = \frac{R_1 I}{R_2 I} = \frac{R_1}{R_2} = \frac{1}{4} \quad \text{or} \quad V_1 : V_2 = 1 : 4$$

Example 9.6
A batch of nominally 100Ω resistors are manufactured with a $\pm 10\%$ tolerance in their resistance values. This means that the resistance of any resistor in the batch may be between $91\,\Omega$ and $110\,\Omega$. If 3 resistors are chosen at random from the batch and connected in series, calculate the maximum and minimum values which could be expected for this series connection.

Solution
The resistance of 3 resistors with resistances R_1, R_2, R_3 connected in series is

$$R = R_1 + R_2 + R_3$$

If the resistors were all to have the maximum permitted value, then

$$R_1 = R_2 = R_3 = 110\,\Omega$$

and so

$$R = 3 \times 110 = 330\,\Omega$$

If the resistors were all at the minimum value of $91\,\Omega$, then

$$R = 3 \times 91 = 273\,\Omega$$

Example 9.7
Determine the current I and the voltages across the individual resistors in the series circuit of fig. 9.23.

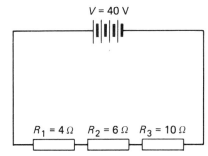

Fig. 9.23

Solution
The total resistance of the circuit is

$$R = R_1 + R_2 + R_3 = 4 + 6 + 10 = 20\,\Omega$$

and the current,

$$I = \frac{V}{R} = \frac{40}{20} = 2\,\text{A}$$

The voltage across R_1,

$$V_1 = R_1 I = 4 \times 2 = 8\,\text{V}$$

The voltage across R_2,

$$V_2 = R_2 I = 6 \times 2 = 12\,\text{V}$$

The voltage across R_3,

$$V_3 = R_3 I = 10 \times 2 = 20\,\text{V}$$

Check: The voltages across the 3 resistors must together equal the applied e.m.f.: $8 + 12 + 20 = 40\,\text{V}$, so our calculations appear correct.

9.13
Properties of parallel circuits: (a) the sum of the currents in the resistors is equal to the total current flowing in the network; (b) the potential difference is the same across the resistors

(a) The sum of the currents in resistors connected in parallel is equal to the current flowing into the parallel network

Figure 9.24(a) shows a circuit in which the two resistors R_1 and R_2 are connected in parallel. The total current I supplied by the battery equals the sum of the respective currents I_1 and I_2 flowing in R_1 and R_2, i.e. $I = I_1 + I_2$. If this were not true current would be escaping from the circuit and flowing to somewhere else. We know that this is not the case since the current in the circuit is confined to the conductor wires and resistors and

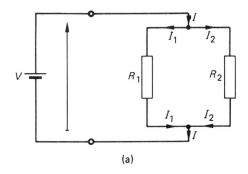

(a)

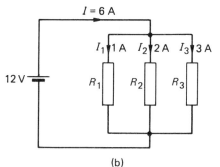

(b)

Fig. 9.24

does not flow into the air, air acting as an insulator (non-conductor of electricity).

In general the sum of the individual currents in a number of resistors connected in parallel is equal to the current flowing into the parallel network, where the term network in this case is used to describe the combination of resistors in parallel. More generally a network is defined as a connection of circuit components. For example, we can describe the connection of two or more resistors in series as a series network; and in fig. 9.24 the connections of the two and the three resistors in parallel as parallel networks. For the parallel network of fig. 9.24(b) we have

$$I = I_1 + I_2 + I_3$$

i.e.

$$6 = 1 + 2 + 3\,\text{A}$$

(b) The potential difference (voltage) is the same across resistors in parallel

The voltage across each resistor in a parallel network is the same. Thus in fig. 9.24(a) the voltage across resistor R_1 equals the voltage across resistor R_2 which is equal to the battery voltage V. Note the important assumption, always made in circuit diagrams, that there is zero potential drop along connecting wires. Any connecting wire drawn in on a circuit diagram is always assumed to have zero resistance and therefore zero voltage drop along it.

In fig. 9.24(b) the voltage across the resistors R_1, R_2, and R_3 is the same and equal to 12 V. Thus, using this result and given the currents in the resistors, we can calculate the resistance values as follows:

$$R_1 = \frac{V}{I_1} = \frac{12}{1} = 12\,\Omega \quad (\text{as } I_1 = 1\,\text{A})$$

$$R_2 = \frac{V}{I_2} = \frac{12}{2} = 6\,\Omega \quad (\text{as } I_2 = 2\,\text{A})$$

$$R_3 = \frac{V}{I_3} = \frac{12}{3} = 4\,\Omega \quad (\text{as } I_3 = 3\,\text{A})$$

9.14
Derivation of the formula, $1/R = 1/R_1 + 1/R_2 + \cdots$ for the equivalent resistance of resistors connected in parallel

In the last two sub-sections we learnt two very

important properties about the parallel connection of resistors:

(1) The total current equals the sum of the individual currents.
(2) The voltage is the same across resistors connected in parallel.

We shall now apply these two properties to determine the equivalent resistance of networks containing two and three resistors in parallel. In the 2-resistor parallel circuit of fig. 9.25(a), the current through the respective resistors is given by

$$I_1 = \frac{V}{R_1} \quad \text{and} \quad I_2 = \frac{V}{R_2}$$

where V = the applied voltage across the resistors.

Since the total current supplied by the source,

$$I = I_1 + I_2$$

we have

$$I = \frac{V}{R_1} + \frac{V}{R_2} = V\left(\frac{1}{R_1} + \frac{1}{R_2}\right)$$

On comparing the last equation with

$$I = \frac{V}{R} = V\left(\frac{1}{R}\right)$$

where R is the equivalent resistance of the parallel combination of R_1 and R_2, we have

$$\frac{1}{R} = \frac{1}{R_1} + \frac{1}{R_2}$$

or

$$\frac{1}{R} = \frac{R_2 + R_1}{R_1 R_2}$$

giving

$$R = \frac{R_1 R_2}{R_1 + R_2}$$

e.g. if $R_1 = 6\,\Omega$, $R_2 = 30\,\Omega$, then

$$\frac{1}{R} = \frac{1}{R_1} + \frac{1}{R_2} = \frac{1}{6} + \frac{1}{30} = \frac{5+1}{30} = \frac{6}{30}$$

so

$$R = \frac{30}{6} = 5\,\Omega$$

or directly,

$$R = \frac{R_1 R_2}{R_1 + R_2} = \frac{6 \times 30}{6 + 30} = \frac{180}{36} = 5\,\Omega.$$

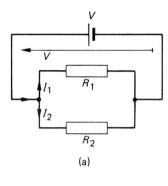

(a)

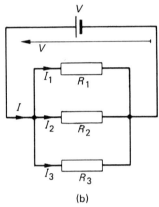

(b)

Fig. 9.25

In the 3-resistor parallel circuit of fig. 9.25(b), the currents through the resistors are

$$I_1 = \frac{V}{R_1}, \quad I_2 = \frac{V}{R_2}, \quad I_3 = \frac{V}{R_3}$$

and since the total current,

$$I = I_1 + I_2 + I_3$$

we have

$$I = \frac{V}{R_1} + \frac{V}{R_2} + \frac{V}{R_3}$$

$$= V\left(\frac{1}{R_1} + \frac{1}{R_2} + \frac{1}{R_3}\right) \qquad (1)$$

but the equivalent resistance representing the effect of R_1, R_2, R_3 in parallel produces the same current I with the same applied voltage V, i.e.

$$I = \frac{V}{R} \qquad (2)$$

so on comparing expressions (1) and (2) we have

$$\frac{1}{R} = \frac{1}{R_1} + \frac{1}{R_2} + \frac{1}{R_3}$$

e.g. (1) If $R_1 = R_2 = R_3 = 12\,\Omega$

$$\frac{1}{R} = \frac{1}{12} + \frac{1}{12} + \frac{1}{12} = \frac{3}{12}$$

so

$$R = \frac{12}{3} = 4\,\Omega$$

e.g. (2) If $R_1 = 18\,\Omega$, $R_2 = 45\,\Omega$, $R_3 = 30\,\Omega$

$$\frac{1}{R} = \frac{1}{18} + \frac{1}{45} + \frac{1}{30} = \frac{5+2+3}{90} = \frac{10}{90}$$

so

$$R = \frac{90}{10} = 9\,\Omega$$

9.15
The solution of problems involving parallel and series-parallel connections of resistors

We solve parallel resistor problems using:

(a) The sum of the currents in resistors connected in parallel equals the current flowing into the parallel network.
(b) The voltage is the same across resistors in parallel.
(c) The equivalent resistance R of

(i) 2-resistors R_1 and R_2 connected in parallel may be found from

$$\frac{1}{R} = \frac{1}{R_1} + \frac{1}{R_2} \quad \text{or} \quad R = \frac{R_1 R_2}{R_1 + R_2}$$

(ii) 3-resistors R_1, R_2, and R_3 in parallel may be found from

$$\frac{1}{R} = \frac{1}{R_1} + \frac{1}{R_2} + \frac{1}{R_3}$$

Remember for series connections,

$$R = R_1 + R_2 + R_3 + \cdots$$

To find the total resistance of a series-parallel connection of resistors:

(i) Find the equivalent resistance of those resistors connected in parallel and
(ii) add it to the total resistance of those resistors connected in series.

Example 9.8
(a) Determine the currents I_1 and I_2 in R_1 and

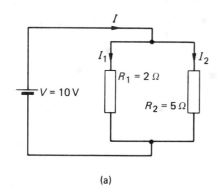

(a)

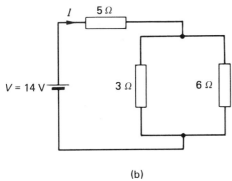

(b)

Fig. 9.26

R_2 and the total current I in the parallel circuit of fig. 9.26(a).
(b) Determine the total resistance R and the total current I in the circuit of fig. 9.26(b).

Solution
(a) Since the voltage applied across R_1 and R_2 is $V = 10$ V,

$$I_1 = \frac{V}{R_1} = \frac{10}{2} = 5\,\text{A}$$

$$I_2 = \frac{V}{R_2} = \frac{10}{5} = 2\,\text{A}$$

The total current I supplied by the battery equals the sum of the current in R_1 and R_2, thus

$$I = I_1 + I_2 = 2 + 5 = 7\,\text{A}$$

(b) The resistance of $3\,\Omega$ and $6\,\Omega$ in parallel is given by

$$\frac{1}{R'} = \frac{1}{3} + \frac{1}{6} = \frac{2+1}{6} = \frac{3}{6} \quad \text{so} \quad R' = \frac{6}{3} = 2\,\Omega$$

is the equivalent resistance in series with $5\,\Omega$.

Thus the total resistance of the circuit is

$$R = 5 + 2 = 7\,\Omega$$

and the total current,

$$I = V/R = 14/7 = 2\,\text{A}$$

Example 9.9
Three resistors of values $R_1 = 30\,\Omega$, $R_2 = 15\,\Omega$, $R_3 = 10\,\Omega$ are connected in parallel. Calculate the equivalent resistance of the combination. If a 10-V battery is connected in series with an ammeter and the 3-paralleled resistors to form a complete circuit, calculate (a) the reading of the ammeter, (b) the currents in R_1, R_2, and R_3.

Solution
The equivalent resistance R of the parallel combination is found from

$$\frac{1}{R} = \frac{1}{R_1} + \frac{1}{R_2} + \frac{1}{R_3} = \frac{1}{30} + \frac{1}{15} + \frac{1}{10}$$

$$= \frac{1+2+3}{30} = \frac{6}{30}$$

so

$$R = \frac{30}{6} = 5\,\Omega$$

The circuit diagram for the second part of the question is drawn in fig. 9.27. The ammeter reads the total circuit current I which is given by

$$I = \frac{V}{R} = \frac{10}{5} = 2\,\text{A}$$

since the battery voltage $V = 10$ V is the voltage applied across the three resistors and $R = 5\,\Omega$ is the equivalent resistance of the three resistors in

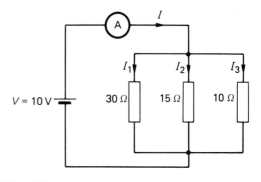

Fig. 9.27

parallel. The respective currents I_1, I_2, I_3 in R_1, R_2, R_3 are

$$I_1 = \frac{V}{R_1} = \frac{10}{30} = \tfrac{1}{3} \text{ A}$$

$$I_2 = \frac{V}{R_2} = \frac{10}{15} = \tfrac{2}{3} \text{ A}$$

$$I_3 = \frac{V}{R_3} = \frac{10}{10} = 1 \text{ A}$$

9.16
Some practical work suggestions

The setting-up of simple parallel and series circuits given a circuit diagram and the measurement of current and voltage

This learning objective is achieved by practical work. The following experiments are examples of the work that could be carried out.

Experiment 1: Measurement of the current versus voltage graph of a linear resistor
Set up a circuit using the circuit diagram of

fig. 9.10(c) as your main guideline. Some suggestions for the individual components are:

battery: three 2-V accumulators or four 1.5 V heavy duty dry battery cells joined in series, i.e. positive terminal of one cell joined to negative of next;

the resistor: a wire-wound resistor in the 20–100 Ω range;

the variable resistor: a rheostat or resistance box providing a variation in resistance from a few tens to several hundred ohms;

voltmeter: capable of measuring voltage in the 0–6 V range;

ammeter: capable of measuring current in the 0–500 mA range.

Note that both the ammeter and voltmeter are fragile instruments and can be permanently damaged by mechanical shock or by passing too great a current through them. It is therefore essential to select a meter (or meter range switch) for the range of current or voltage expected in the circuit, and to connect it in the way explained in Section 9.7.

Draw up a table of V and I values and plot a graph of I versus V as described in Section 9.8(a).

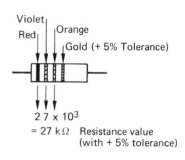

2 7 x 10³
= 27 kΩ Resistance value
(with + 5% tolerance)

(b) Coding of a 27 kΩ resistor

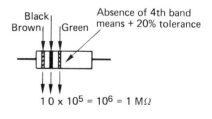

1 0 x 10⁵ = 10⁶ = 1 MΩ

(c) Coding of a 1MΩ resistor

Colour of band	First band	Second band	Third band	Fourth band (tolerance)	
Black	0	0	x 10⁰ = 1	None	± 20%
Brown	1	1	x 10¹	Silver	± 10%
Red	2	2	x 10²	Gold	± 5%
Orange	3	3	x 10³	Red	± 2%
Yellow	4	4	x 10⁴		
Green	5	5	x 10⁵		
Blue	6	6	x 10⁶		
Violet	7	7			
Grey	8	8			
White	9	9			

(a) Colour coding of fixed resistors

Fig. 9.28

Experiment 2: Measurement of the current versus voltage graph of a lamp
Set up the circuit shown in fig. 9.12. Suggestions for the individual components are:

the lamp bulb: conventional 6.5-V, 0.3-A torch lamp bulb mounted, if possible, in a lamp bulb holder so as to make circuit connections easier;

the voltage source: ideally a variable d.c. voltage supply capable of being varied from 0–9 V and delivering up to 0.5 A;

alternatively 2-V accumulators in series or dry cells in series with the inclusion of a variable resistance or rheostat to vary the current and the voltage across the lamp could be used in the same way as in fig. 9.10(c);

ammeter: 0–500 mA; voltmeter 0–10 V.

Measure *V* and *I* points and plot the *I–V* curve of the lamp. Similar results to those given in Section 9.8(b) should be obtained.

Experiment 3: Measurement of resistor values
Resistors used in electronics often have their resistance values indicated on them by a series of coloured bands or rings. The way these bands code the resistance is shown in fig. 9.28. The first two bands give the first two digits (numbers) of the resistance; the third band gives the power of ten by which the first two digits must be multiplied; the fourth band denotes the tolerance of the resistance value, i.e. the maximum percentage error that the resistance may have.

A simple practical set-up and the corresponding circuit diagram to measure the values of resistor components are shown in fig. 9.29. A resistor (or series or parallel combinations of resistors) is connected between two crocodile clips in series with a 1.5 V battery and 0–100-mA ammeter. Thus the maximum current in the circuit should be below 100 mA, which in turn means the minimum value of resistance we can safely measure is $R_{min} = 1.5/0.1 = 15\,\Omega$ with a 0–100 mA meter. A voltmeter is connected across the resistor and this will read approximately 1.5 V, since the battery e.m.f. is quoted nominally at this value.

A useful experiment could be carried out as follows, assuming that we have a suitable supply of resistors. For example if we were provided with 33 Ω and 47 Ω nominal value resistors of

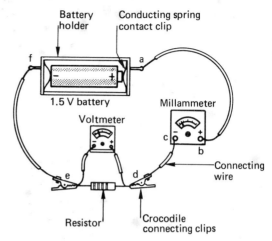

(a) A simple practical set-up to measure resistance values

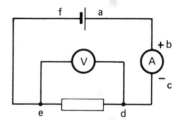

(b) Circuit diagram of practical set-up

Fig. 9.29

10 % tolerance (fourth band silver):

33 Ω: first band orange, second band orange, third band black.
47 Ω: first band yellow, second band violet, third band black.

(1) Connect each resistor in turn in the circuit between the crocodile clips and record the current *I* and voltage *V* readings on the respective meters. Tabulate these readings and calculate the individual resistor values. Check that these values are within the 10 % tolerance limit, e.g. suppose we recorded the following results for a nominal 47-Ω resistor:

$$I = 31\,\text{mA}, \quad V = 1.40\,\text{V}$$

then

$$R = \frac{V}{I} = \frac{1.4}{0.031} = 45.2\,\Omega$$

10 % of 47 Ω = 47 × 10/100 = 4.7 Ω, so the resistor should have a resistance within the range

$47 - 4.7 = 42.3\,\Omega$ and $47 + 4.7 = 51.7\,\Omega$. Clearly in the above example where $R = 45\,\Omega$ the resistor is well within the 10% tolerance limits.

(2) Make up series combinations of

$$33\,\Omega + 33\,\Omega, \quad 47\,\Omega + 47\,\Omega, \quad 33\,\Omega + 47\,\Omega,$$
$$33\,\Omega + 33\,\Omega + 47\,\Omega.$$

Insert these combinations between the crocodile clips one at a time. Record I and V and hence calculate the resistance of each combination. Check that these resistances agree with theory, i.e. $R = R_1 + R_2$ or $R_1 + R_2 + R_3$.

(3) Make up parallel combinations of

$$33\,\Omega + 33\,\Omega, \quad 47\,\Omega + 47\,\Omega, \quad 47\,\Omega + 47\,\Omega + 47\,\Omega.$$

Insert these combinations between the crocodile clips in turn. Record I and V and hence calculate the resistance of each combination. Check that these resistances agree with theory, i.e.

$$\frac{1}{R} = \frac{1}{R_1} + \frac{1}{R_2} \quad \text{or} \quad \frac{1}{R} = \frac{1}{R_1} + \frac{1}{R_2} + \frac{1}{R_3}$$

9.17
The relationship between the resistance of a conductor and its length, cross-sectional area and material

The resistance of a linear conductor (i.e. one which obeys Ohm's Law) depends upon three factors:

(1) The cross-sectional area of the conductor.
(2) The length of the conductor.
(3) The material from which the conductor is made.

The last property is measured by the resistivity of the material, which we shall define in the next section. At the moment we shall say the resistivity will be denoted by the Greek symbol ρ pronounced 'rho', and ρ has a numerical value which varies from material to material. For good conductors like copper and aluminium ρ is an extremely small number. For conductors which are used to make, for example, wire-wound resistors ρ will be several tens of times larger.

The resistance of a uniform linear conductor, such as a metallic wire of constant cross-sectional area, is given by

$$R = \rho\,\frac{l}{A}$$

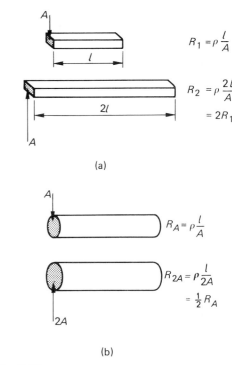

(a)

(b)

Fig. 9.30

where l is the conductor length in metres,
$\quad A$ is the cross-sectional area in square metres,
$\quad \rho$ is the resistivity of the conductor material.

Thus if we have a conductor of a given material and this conductor has a constant cross-sectional area, its resistance will be directly proportional to its length. This means, for example, if we double the length of the conductor, we double its resistance. This situation is illustrated in fig. 9.30(a). If, however, we double the cross-sectional area of a conductor but keep its length fixed, then the resistance is halved. This effect occurs because the resistance is inversely proportional to cross-sectional area, *see* fig. 9.30(b).

9.18
The definition of resistivity

The resistivity ρ of a conductor material at a given temperature is defined as the resistance between the opposite faces of a one-metre cube of the material when the current flows perpendicularly between two of its faces as shown in

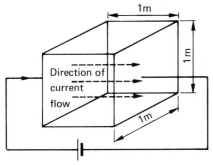

Fig. 9.31

fig. 9.31. Since the resistance of a material depends on temperature, its resistivity should be quoted at a given temperature.

The units of resistivity are ohm-metres (Ωm).

This may be seen by analysing the units of the terms in the above expression for R:

$$R = \rho \frac{l}{A} \quad \text{so} \quad \rho = R \frac{A}{l}$$

and the units of

$$\rho = (\text{ohms}) \times \frac{\text{square metres}}{\text{metres}} = \Omega \times \frac{\text{m}^2}{\text{m}} = \Omega\text{m}.$$

Tables 9.4 and 9.5 list the resistivities of some commonly used conducting and insulating materials.

Copper, followed by aluminium, is the most widely used metal for the conduction of electricity for both industrial and domestic applications. Copper has a lower resistance for a

Table 9.4 *Resistivity and temperature coefficient values for conductors*

Material	Resistivity ρ (ohm-metres at 20°C)	Relative resistance $= \dfrac{\rho \text{ material}}{\rho \text{ copper}}$	Temperature coefficient of resistance α (per °C, over range 0–100°C)
copper	1.72×10^{-8}	1	0.0039
aluminium	2.8×10^{-8}	1.64	0.0039
iron	9.6×10^{-8}	5.6	0.006
platinum	10.6×10^{-8}	6.16	0.0039
silver	1.6×10^{-8}	0.95	0.0038
gold	2.4×10^{-8}	1.42	0.0034
steel	12 to 90×10^{-8}	7 to 53	varies considerably
carbon	$\sim 7000 \times 10^{-8}$	~ 4000	-0.0005
manganin	45×10^{-8}	26	$\pm 0.000\,02$
constantan	49×10^{-8}	28.5	$\pm 0.000\,2$

Table 9.5 *Resistivity values for some common insulators*

Material	Resistivity ρ (ohm-metres at 25°C)	Dielectric strength (volts per metre)
ceramics (magnesium silicate composition)	$> 10^{12}$	$> 7 \times 10^6$
porcelain	$> 10^{11}$	$> 3 \times 10^6$
glass	$> 10^{7}$	
bakelite	$> 10^{9}$	$> 8 \times 10^6$
polythene	$> 10^{15}$	$> 10 \times 10^6$
paper	$> 10^{9}$	$> 10 \times 10^6$
mica	$> 10^{11}$	$> 100 \times 10^6$
rubbers	$> 10^{11}$	$> 100 \times 10^6$
distilled water	$\sim 10^{6}$	
air (atmospheric pressure)		3×10^6

(N.B. $>$ sign means greater than. Dielectric strength defines, in volts per metre thickness of the material, the voltage which an insulator can withstand without breaking down its non-conducting properties.)

given cross-section whilst aluminium has a lower resistance for a given weight. Increased mechanical strength may be obtained by using composite conducting materials. For example, steel-cored aluminium and copper-clad steel are used, the steel supplying the strength and the aluminium and copper providing the good electrical conducting properties. Such properties are required by overhead lines used to distribute electricity. These conducting lines must have sufficient mechanical strength to hang freely in a fairly long span and save the number of pylons required. They must also possess low resistance to carrying full current without undue heat loss or temperature rise. Forms of carbon are also used to conduct electricity; for example, carbon brushes in an electrical motor and carbon resistors. Precious materials, such as platinum and gold, are sometimes used for specialist applications where a metal such as copper may deteriorate through oxidisation.

Insulating materials do not conduct electricity and are therefore used for covering or separating conductors carrying current, or to separate conductors at different voltages to prevent current flow between the conductors. For example, ceramic materials are used for constructing insulators to carry conductors for electrical supply and lines for telecommunication applications. Plastics are widely used as insulators in electrical appliances and electronic equipment. Plastic material, such as polythene, is also used to separate conductors in wiring cable (for example, in two- and three-wire lighting and power cable) and coaxial cables for telecommunications purposes. An example of the latter is the coaxial cable which joins an aerial to a television set. The inner central wire is separated from the outer conducting sheath by polythene.

9.19
The solution of problems involving resistivity

Remember,

$$R = \rho \frac{l}{A}$$

where ρ = resistivity; units, ohm-metres (Ωm);
$\quad l$ = length of conductor; units, metres (m);
$\quad A$ = cross-sectional area of conductor; units, square metres (m²).

Example 9.10
The resistance of 1 m length of constantan resistance wire of constant cross-section is 3.5 Ω. Calculate:

(a) The resistance of a resistor consisting of 20 m of the above wire.
(b) The resistance of 1 m of constantan wire if its cross-sectional area is (i) four times that of the 3.5 ohm per metre wire, (ii) half that of the 3.5 ohm per metre wire.

Solution
(a) Since the resistance of a uniform conductor is proportional to its length, the resistance R of 20 m of 3.5 Ω/m wire is

$$R = 3.5 \times 20 = 70 \, \Omega$$

(b) Since the resistance of a uniform conductor is inversely proportional to its cross-sectional area,

(i) increasing the area by $\times 4$ will reduce the resistance by a factor of 4, so the resistance of 1 m = 3.5 ÷ 4 = 0.875 Ω.
(ii) reducing the area by half will increase the resistance by a factor 2, so the resistance of 1 m = 3.5 $\times$ 2 = 7.0 Ω.

Example 9.11
Determine the resistance at 0 °C of a coil made from a 500 m length of wire of radius 0.46 mm. The resistivity of the wire material is $\rho = 1.56 \times 10^{-8} \, \Omega$m at 0 °C. Take $\pi = 3.142$.

Solution
The radius of the wire $r = 0.46$ mm $= 0.46 \times 10^{-3}$ m, the cross-sectional area of the wire,

$$A = \pi r^2$$
$$= 3.142 \times (0.46 \times 10^{-3})^2 = 0.665 \times 10^{-6} \, \text{m}^2$$

Hence the resistance of the coil of length $l = 500$ m at 0 °C is

$$R = \rho \frac{l}{A} = \frac{1.56 \times 10^{-8} \times 500}{0.665 \times 10^{-6}}$$

$$= \frac{1.56 \times 5}{0.665} = 11.73 \, \Omega$$

Example 9.12
The resistance at 20 °C of a 1 km length of conductor rail of uniform cross-sectional area 5600 mm² was measured at 0.025 Ω. Calculate its resistivity at 20 °C.

If a cubic metre of the same material as the conductor rail were drawn out into a wire of 1 mm radius, find the total resistance of the wire so formed, at 20 °C. Take $\pi = 3.142$.

Solution
Using the formula,

$$R = \rho \frac{l}{A}$$

we have for the resistivity,

$$\rho = R \frac{A}{l}$$

and on substituting $R = 0.025\,\Omega$,

$$A = 5600\,\text{mm}^2 = 5600 \times 10^{-6}\,\text{m}^2,$$
$$l = 1\,\text{km} = 10^3\,\text{m}$$

we have (at 20 °C),

$$\rho = \frac{0.025 \times 5600 \times 10^{-6}}{10^3}$$

$$= 0.025 \times 5600 \times 10^{-9} = 140 \times 10^{-9}\,\Omega\,\text{m}$$

Let l be the length of wire obtained from $1\,\text{m}^3$. Then, as the wire has a radius $r = 1\,\text{mm} = 10^{-3}\,\text{m}$, its cross-sectional area $A = \pi r^2 = \pi \times 10^{-6}\,\text{m}^2$. The volume of the wire $= 1\,\text{m}^3 = Al\,\text{m}^3$ so

$$l = \frac{1}{A} = \frac{1}{\pi \times 10^{-6}} = \frac{10^6}{\pi} = \frac{10^6}{3.142}$$

$$= 3.183 \times 10^5\,\text{m}$$

Hence the total resistance of the wire at 20 °C is

$$R = \rho \frac{l}{A} = \frac{140 \times 10^{-9} \times 3.183 \times 10^5}{3.142 \times 10^{-6}}$$

$$= \frac{140 \times 3.183}{3.142} \times 10^2 = 14.18\,\text{k}\Omega$$

9.20
The resistance of a conductor varies with temperature

The resistance of a conductor varies with temperature. The resistance of most metal and alloy conductors increases with a rise in temperature. However, not all conductors have a resistance that increases with temperature. Carbon, most semi-conductor materials, and

electrolytic solutions have a resistance which decreases as temperature increases.

Over a limited temperature range, for example between 0° and 100 °C, the resistance of a conductor is given quite accurately by the formula

$$R_T = R_0(1 + \alpha T)$$

where R_T = resistance of the conductor at T °C
R_0 = resistance of the conductor at 0 °C

and α is a constant known as the temperature coefficient of resistance.

Using the above equation to make α the subject of the formula, we have

$$R_T = R_0 + R_0 \alpha T$$
$$R_T - R_0 = R_0 \alpha T$$

so

$$\alpha = \frac{R_T - R_0}{R_0 T}$$

This expression defines the temperature coefficient of resistance α. In words we may define the temperature coefficient of resistance of a conductor as the fractional increase in the resistance of a conductor at 0 °C per degree celsius rise in temperature. Values of α for some common conducting materials are given in Table 9.4. Note that the values of α for manganin and constantan are extremely small compared with the other conducting materials. These low values of α make manganin and constantan attractive materials for producing wire-wound standard resistors, since resistance changes with temperature will be minimised. The variation of resistance with temperature of platinum, however, is used in the platinum resistance thermometer for the accurate determination of temperature over the range -150 °C to above 1000 °C.

Example 9.13
(a) A long length of copper wire wound in the form of a coil has a resistance of $50\,\Omega$ at 0 °C. Given that the temperature coefficient of resistance of copper $\alpha = 0.0043$ per degree celsius, calculate its resistance at 20 °C.
(b) Calculate the resistance of a carbon resistor at 75 °C if its resistance at 0 °C is $100\,\Omega$ and $\alpha = -0.0005$ per °C.

Solution
(a) The resistance of the coil at 0 °C, $R_0 = 50\,\Omega$

and since $\alpha = 0.0043$ per °C, the resistance at 20°C is

$$R_{20} = R_0(1 + \alpha 20)$$
$$= 50(1 + 20 \times 0.0043) = 54.3\,\Omega$$

(b) Since $R_0 = 100\,\Omega$, $\alpha = -0.0005$ per °C, the resistance of the resistor at 75°C is

$$R_{25} = 100(1 - 0.0005 \times 75) = 96.2\,\Omega$$

Example 9.14

A temperature-sensitive resistor element may be used to measure temperature indirectly through measurement of its resistance. If the resistance of such an element at 0°C is 50 Ω and its temperature coefficient of resistance $\alpha = 0.0039/$°C, calculate the temperature at which its resistance is (i) 61.7 Ω, (ii) 46.7 Ω. Assume α remains constant over temperature range -20°C to $+80$°C.

Solution

Solving $R_T = R_0(1 + \alpha T)$ for temperature T we have,

$$R_T = R_0 + R_0\alpha T$$
$$R_T - R_0 = R_0\alpha T$$

so

$$T = \frac{R_T - R_0}{R_0\alpha}$$

(i)   When $R_T = 61.7\,\Omega$:

$$T = \frac{61.7 - 50}{50 \times 0.0039} = \frac{11.7}{0.195} = 60\,°C$$

(ii) When $R_T = 46.7\,\Omega$:

$$T = \frac{46.7 - 50}{50 \times 0.0039} = \frac{-3.3}{0.195} = -16.9\,°C$$

9.21

The comparison of the merits of wiring lamps in series and in parallel

Figure 9.32(a) shows a circuit in which three lamps are wired in series. The same current will pass through each of the lamps and a voltage will be developed across each depending on their respective values of resistance. The sum of these voltages equals the applied battery voltage, i.e.

$$V = V_1 + V_2 + V_3$$

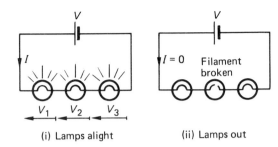

(i) Lamps alight   (ii) Lamps out

(a) Series connection of lamps

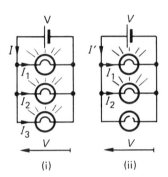

(b) Parallel connection of lamps

Fig. 9.32

Unfortunately if any one of the lamps breaks down, for example, if its filament which conducts the current fractures, this will cause a break in the circuit. Thus current can no longer flow and all the lamps will go out even though a fault only occurs in one. Such a situation is illustrated in fig. 9.32(a)(ii).

The parallel wired circuit, shown in fig. 9.32(b), has the important merit that if a lamp does break down a continuous circuit is still maintained by the other lamps in parallel. In a parallel circuit each lamp has the same voltage across it and the sum of the individual lamp currents equals the total current supplied by the source, i.e.

for fig. (i)

$$I = I_1 + I_2 + I_3;$$

for fig. (ii)

$$I' = I_1 + I_2$$

One important merit of series wiring of lamps is the case when we have a number of low-voltage lamps to be operated from a fixed high-voltage supply. For example, in the lighting of a

Christmas tree we may have 40 or so low-voltage lamps and we would normally wish to operate them from a mains supply which has an effective voltage of 240 V. To connect these lamps in parallel would mean that 240 V existed across each lamp and apart from each lamp consuming a great deal of power and getting dangerously hot, a low-voltage lamp (e.g. a 6 V rated one) would not be designed to cope with such high voltages. In such an application it is obvious that series wiring (or perhaps a combination of series and some parallel wiring) should be used. House lighting bulbs, however, are specifically designed to operate at 240 V and are therefore wired in parallel. It would be disastrous to wire in series, since with one break in a series circuit all lights would go out. In our Christmas tree lighting such a fault can be tolerated and rectified.

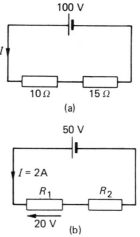

Fig. 9.33   for Problem 2

Problems 9:   Resistive electrical circuits

1.   (a) Write down the SI prefixes for $\times 10^9$, $\times 10^6$, $\times 10^3$, $\times 10^{-3}$, and $\times 10^{-6}$. Use the appropriate prefix to express the following concisely: 0.000 012 V, 5 000 000 Ω, 0.007 A.
(b) Draw the preferred circuit symbols for a battery, a resistor, a variable resistor, an ammeter, a voltmeter, a lamp, and a switch.
(c) Draw the circuit diagram of a battery in series with a switch and two resistors. Show on this diagram the positions of an ammeter to measure the circuit currents and a voltmeter to measure the voltage across the battery terminals.
(d) Sketch graphs showing the $V-I$ relationship between the voltage across one of the resistors and the current in the circuit described in (c), when the resistor is (i) a wire-wound (linear) resistor, (ii) a lamp bulb (non-linear) resistor.
2.   (a) Calculate the current I and the voltage across the 15-Ω resistor in the series circuit of fig. 9.33(a).
(b) The current in the series circuit of fig. 9.33(b) is $I = 2$ A. The source voltage is 50 V and the voltage across R_1 is 20 V. Calculate the voltage across R_2 and the values of R_1 and R_2.
3.   (a) State Ohm's Law. Give two examples of components which do not obey Ohm's Law.
(b) Two linear resistors, $R_1 = 60\,\Omega$ and $R_2 = 30\,\Omega$, are connected in series with a 90-V battery. Draw the circuit diagram and calculate the current in the circuit and the voltage across the 30-Ω resistor.

(c) R_1 and R_2 are now connected in parallel and the battery placed across their terminals. Draw the circuit diagram and calculate the current delivered by the battery and the current in the 60-Ω resistor.
4.   Calculate the current and the voltage across each of the resistors in the series circuit of fig. 9.34.
If the resistors have a $\pm 10\%$ tolerance in their resistance values, calculate the maximum and the minimum values of current that could occur in the circuit.

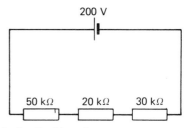

Fig. 9.34   for Problem 4

5.   (a) Calculate the currents I_1 and I_2 and the total current I in the parallel circuit of fig. 9.35.
(b) Calculate the total equivalent resistance and the current supplied by the battery in the circuit of fig. 9.35(b).
6.   Three resistors of values $R_1 = 20\,\Omega$, $R_2 = 30\,\Omega$, $R_3 = 60\,\Omega$ are connected in parallel. Calculate the equivalent resistance of this combination. A 20-V battery is connected in series with an ammeter and the paralleled resistors to form a complete circuit. Draw the circuit diagram and calculate the expected reading on the ammeter and the currents in each of the resistors.

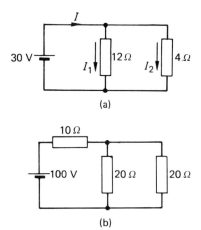

(a)

(b)

Fig. 9.35 for Problem 5

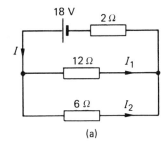

(a)

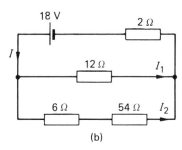

(b)

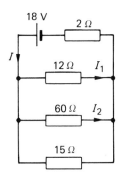

(c)

Fig. 9.36 for Problem 7

7. Calculate the currents I, I_1, I_2 in the circuits of fig. 9.36.

8. State four factors on which the resistance of a metallic conductor depends.

The resistance of a metre length of constantan wire of a given cross-section is $4\,\Omega$. Calculate:

(a) The resistance of a resistor consisting of 25 m of this wire.

(b) The resistance of constantan wire in ohms per metre if its cross-sectional area is (i) half, (ii) double, that of the $4\text{-}\Omega/\text{m}$ wire.

9. Define the electrical resistivity of a material. Give three examples of good electrically conducting materials and three examples of insulators in common use.

Determine the resistance of a coil of wire of length 200 m and cross-sectional area $1\,\text{mm}^2$. The resistivity of the wire material is $50 \times 10^{-8}\,\Omega\text{m}$.

10. Calculate the resistance of a 100 m reel of copper wire of 0.25 mm radius at 20 °C, 0 °C, 40 °C, given that the resistivity of copper is $1.74 \times 10^{-8}\,\Omega\text{m}$ at 20 °C and the temperature coefficient of resistance of copper is $0.004\,°\text{C}^{-1}$.

11. Define the temperature coefficient of resistance of a material and state the formula for the resistance of a conductor at $T\,°\text{C}$.

The resistance of a wire-wound resistor is $108\,\Omega$ when measured at 20 °C. When the resistor is immersed in melting ice at 0 °C its resistance is $100\,\Omega$. Calculate the temperature coefficient of resistance of the resistance wire and the resistance of the resistor at 45 °C.

12. The current in a circuit consisting of a 30 V source in series with a resistor is 150 mA when the resistor is at 0 °C, and 129 mA when the resistor is at 20 °C. Calculate:

(a) the resistance of the resistor at 0 °C and 20 °C,

(b) its temperature coefficient of resistance, (c) its resistance at $-10\,°\text{C}$ and at 50 °C.

13. Draw circuit diagrams showing the wiring of six lamps in series, in parallel, and in series-parallel. Compare the relative merits of wiring lamps in series and in parallel.

14. The resistance of a conductor is $104\,\Omega$ at 10 °C and $118\,\Omega$ at 40 °C. Determine the temperature coefficient of resistance of the conductor and its resistance at 0 °C.

15. The current in a circuit consisting of a 10 V source in series with a resistor is 400 mA when measured at 20 °C. Calculate the current in the circuit when the temperature is increased to 35 °C if the temperature coefficient of the resistor material is (i) $\alpha = 0.005/\,°\text{C}$, (ii) $\alpha = 0.0001/\,°\text{C}$.

10 Power in electrical circuits

The expected learning outcome of this chapter is that the student calculates power in electrical circuits.

10.1
The power produced in a resistive circuit is given by
$$P = IV = I^2R = V^2/R \text{ watts}$$

In Section 9.3 we stated that a voltage must exist between two points in a conductor to produce current flow, i.e. to supply energy to transport the charge. We defined the voltage between two points as the work done in transporting a unit charge between those two points.

We shall now apply this definition to find the energy and power supplied by a source to a resistive circuit. In the circuit of fig. 10.1, the

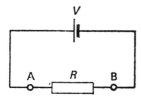

Fig. 10.1

work done in transporting a unit of charge (1 coulomb) through the resistor R from A to B equals the voltage V applied between A and B by the voltage source. The energy supplied by the source to the charge (the charge in this case being the free electrons in the circuit) is used up by doing work to overcome the resistance presented by the conductor. Thus a continuous supply of electrical energy by the source enables a steady current to be established through the conductor, the electrical energy being converted to heat in the conductor.

In the circuit of fig. 10.1, the source will sustain a steady current I which depends on the voltage V between A–B and the conductor resistance R. I can of course be calculated by applying Ohm's Law, i.e. $I = V/R$. Now remembering that for steady current flow:

current = rate of flow charge

= charge flowing per second

or

charge transported = current × time,

the total charge Q transported through R if I flows for t seconds is

$$Q = It \text{ coulombs.}$$

Further, since the energy supplied by the source to transport one coulomb from A to B through R is V, the total energy W supplied by the source to transport a charge Q is given by

$$W = Q \times V$$
$$= (It)V = IVt \text{ joules}$$

Remember that the joule (J) is the unit of work and energy and one volt equals one joule per coulomb.

Power is defined as the rate of supply of energy, or equivalently the rate of doing work, so the power produced in the circuit is equal to the energy supplied per second which in turn is equal to the work done per second against the resistance. Thus, if we use the symbol P for power, the power produced in the circuit is given by

$$P = \frac{W}{t} = IV \text{ watts}$$

The SI unit of power is the watt, symbol W. One watt equals one joule per second, i.e. W = 1 J/s; 1 kW = 1000 J/s.

Although the unit of energy and work is the joule, energy is sometimes expressed in terms of kilowatt hours (kWh),

1 kWh = (1000 watts)

× (no. of seconds in 1 hour)

= 1000 × (60 × 60) = 3.6 MJ

We have seen that power P supplied to and dissipated in a resistive circuit is given by,

$$P = IV \text{ watts}$$

where I = total current flowing in circuit,
V = total voltage across the circuit.

In the circuit of fig. 10.1, we have by Ohm's Law,

$V = RI$ so $P = IV = I(RI) = I^2R$ watts, or alternatively substituting for $I = V/R$,

$$P = \frac{V^2}{R} \text{watts}$$

In general, for any circuit consisting of any combination of resistors:

$$P = IV = I^2R = \frac{V^2}{R} \text{watts}$$

where $I =$ total current flowing into the resistive network,
$V =$ voltage across the resistor network,
$R =$ equivalent resistance of resistor network.

Remember also that the total power produced in a circuit is equal to the sum of the I^2R powers dissipated in each of the individual resistors.

10.2
The calculation of the power dissipated in simple circuits

We shall now apply the results:

power, $P = IV = I^2R = \dfrac{V^2}{R}$ watts

energy, $W = Pt = IVt$ joules

of the last section to calculate the power and energy supplied to a circuit by a source. In a circuit consisting of resistors, this power is dissipated as heat in the resistor components themselves.

Example 10.1
(a) Calculate the power dissipated in the circuit of fig. 10.2(a).

(b) Calculate the powers dissipated in the resistor R_1, in the resistor R_2, and the total power dissipated in the circuit of fig. 10.2(b).
(c) Calculate the power dissipated in resistors R_1 and R_2 and the total power supplied by the source to the circuit of fig. 10.2(c).

Solution
(a) The total resistance of the circuit, $R = 16 + 4 = 20\,\Omega$, hence the power dissipated in the circuit,

$$P = \frac{V^2}{R} = \frac{40^2}{20} = \frac{1600}{20} = 80 \text{ W}$$

(b) The current flowing through R_1,

$$I_1 = \frac{60}{6} = 10 \text{ A},$$

hence the power dissipated in R_1, $P_1 = I_1^2R_1 = 10^2 \times 6 = 600$ W.
The current through R_2,

$$I_2 = \frac{60}{30} = 2 \text{ A},$$

so power dissipated in R_2, $P_2 = I_2^2R_2 = 2^2 \times 30 = 120$ W.
The total power dissipated P in the circuit equals the sum of the individual powers, so

$$P = P_1 + P_2 = 600 + 120 = 720 \text{ W}$$

N.B.: Alternative methods:

$$P_1 = \frac{V^2}{R_1} = \frac{60^2}{6} = 600 \text{ W},$$

$$P_2 = \frac{V^2}{R_2} = \frac{60^2}{30} = 120 \text{ W},$$

$$P = \frac{V^2}{R}$$

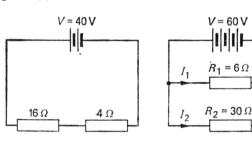

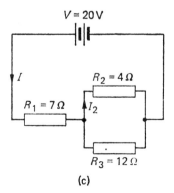

(a)   (b)   (c)

Fig. 10.2

where R is the equivalent resistance of $6\,\Omega$ in parallel with $30\,\Omega$.

Thus

$$\frac{1}{R} = \frac{1}{6} + \frac{1}{30} = \frac{5+1}{30} = \frac{6}{30} = \frac{1}{5},$$

$$P = \frac{60^2}{5} = 720\,\text{W}$$

(c) The resistance R' of R_2 in parallel with R_3 is found from

$$\frac{1}{R'} = \frac{1}{4} + \frac{1}{12} = \frac{3+1}{12} = \frac{4}{12} \quad \text{so} \quad R' = 3\,\Omega$$

Hence the total resistance R of the circuit is

$$R = R_1 + R' = 7 + 3 = 10\,\Omega$$

and so the total current,

$$I = \frac{V}{R} = \frac{20}{10} = 2\,\text{A}$$

The power P_1 dissipated in R_1 is $P_1 = I^2 R_1 = 2^2 \times 7 = 28\,\text{W}$.
The power dissipated in R_2,

$$P_2 = I_2^2 R_2$$

where the current I_2 in R_2 may be found as follows:
the voltage across R_2 (and R_3) is

$$V_2 = R'I = 3 \times 2 = 6\,\text{V}$$

thus $I_2 = V_2/R_2 = 6/4 = 1.5\,\text{A}$,
and hence $P_2 = 1.5^2 \times R_2 = 1.5^2 \times 4 = 9\,\text{W}$.
The total power P supplied by the source is

$$P = IV = 2 \times 20 = 40\,\text{W}$$

Example 10.2
(a) Calculate the current flowing in and the resistance of a heater filament of a car-demister which is rated at 6 W when working from a 12-V battery.
(b) An electric fire takes a current of 8.3 A when connected to 240-V supply. Calculate:

   (i) the power dissipated by the fire,
   (ii) the resistance of the heating element in the fire,
   (iii) the total energy supplied if the fire is switched on for 8 hours,
   (iv) the cost of running the fire for 8 hours if the tariff charged is 2.2 p per kilowatt hour.

Solution
(a) Let the current through the heater filament be I when connected to the $V = 12$ V battery. Then the power dissipated in the filament, given at 6 W, provides the relation:

$$6 = VI = 12I$$

so $I = \frac{6}{12} = 0.5\,\text{A}$ and the resistance R of the filament is

$$R = \frac{V}{I} = \frac{12}{0.5} = 24\,\Omega$$

(b) (i) power $P = IV = 8.3 \times 240 = 1992\,\text{W}$ or 1.992 kW,
   (ii) resistance

$$R = \frac{V}{I} = \frac{240}{8.3} = 28.9\,\Omega,$$

(iii) energy $W = IVt$ joules.
where in this case $t = 8$ hours $= 8 \times 60 \times 60 = 28\,800$ s. Thus $W = 8.3 \times 240 \times 28\,800 = 57.37\,\text{MJ}$,
or in kilowatt hours,

$$\begin{aligned} W &= IV \times (\text{hours}) \div 1000 \\ &= 8.3 \times 240 \times 8 \div 1000 \\ &= 15.94\,\text{kWh} \end{aligned}$$

(iv) The cost of running the fire is (cost per kilowatt hour) × (no. of kWh of energy supplied) $= 2.2 \times 15.94 = 35\,\text{p}$.

10.3
Thermal electrical fuses and the calculation of fuse values given the power rating and voltage of an appliance

Under normal circumstances, when current is passed through a conductor, the temperature of the conductor rises until the heat generated within the conductor is equal to the heat flowing to its surroundings. However, if the current through a conductor is excessive, the heating effect may be sufficient to raise the temperature of the conductor to its melting point, and hence melt the conductor. This principle is used in the common wire fuse, examples of which are shown in fig. 10.3(a). An electrical fuse is used as a safety device to prevent damage which may be otherwise caused by excessive current flow and its associated heating effect. The wire fuse consists of a relatively fine conducting wire which will melt when the current through it exceeds certain

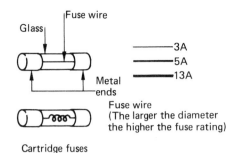

Cartridge fuses

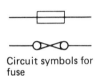

Circuit symbols for fuse

(a) Wire fuses

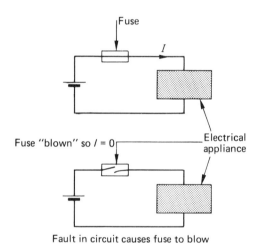

Fuse "blown" so $I = 0$

Fault in circuit causes fuse to blow

(b) Position of fuse in a circuit

Fig. 10.3

values, e.g. common household fuses are rated at 3 A, 5 A, 13 A, and 15 A. Thus, if an electrical appliance is fitted with a fuse in series with the supply, and a fault occurs which causes a current in excess of the rated value of the current which the fuse can safely take, the fuse will melt, causing a break in the circuit. This break will disconnect the supply of the appliance, as shown for example in fig. 10.3(b).

Example 10.3
Determine the value of a wire fuse, rated in amperes, which should be used in a domestic circuit taking a power of 3 kW from 240 V supply.

Solution
The power P taken from the supply,

$$P = IV = 3 \, \text{kW} = 3000 \, \text{W}$$

where $V = 240 \, \text{V}$, the supply voltage, and $I =$ current flowing in the circuit. Thus

$$3000 = I \times 240$$

so

$$I = \frac{3000}{240} = 12.5 \, \text{A}$$

Hence the fuse rating must be at least 12.5 A just to allow the required current flow. To permit minor variations in supply (not fault conditions) a sensible rating value would be 15 A.

Example 10.4
Figure 10.4 shows a circuit which is to be used to charge a car battery. The variable resistor R is included in the circuit to limit the current to below 6 A. A 6 A fuse is included for protection. Calculate the minimum value of R which should be used. The fuse may be assumed to have negligible resistance.

Solution
The circuit equation is set up using Ohm's Law and the series circuit condition:

applied source voltage = voltage across R
   + car battery voltage

so

$$30 = RI + 12$$

and substituting the maximum permitted value of I, i.e. $I = 6 \, \text{A}$ we have,

$$30 = 6R + 12$$

i.e.

$$6R = 30 - 12 = 18$$

$$R = \frac{18}{6} = 3 \, \Omega$$

where R is the minimum value; any lower value will blow the fuse.

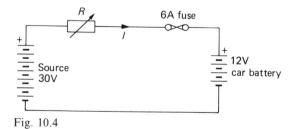

Fig. 10.4

Problems 10:   Power in electrical circuits

1.   Show that the power P dissipated in a resistor of R ohms is given by $P = I^2R$ watts, where I is the current flowing in the resistor in amperes.

An electric fire produces 3 kW of heat when connected to a 240-V supply. Calculate:
(a) The resistance of the heating elements in the fire.
(b) The total energy supplied if the fire is switched on for 4 hours.
(c) The cost of running the fire for 4 hours if the tariff charged is 3 p per kilowatt hour.

2.   Calculate the power dissipated in resistor R_1, in R_2, and in R_3 in the circuit of fig. 10.5.

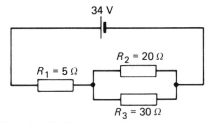

Fig. 10.5   for Problem 2

Calculate also the total power supplied by the source to the circuit.

3.   Calculate the voltage drop across the ends and the total power dissipated in a 100 m length of copper strip of rectangular cross-section of dimensions 10×2 mm, when a current of 40 A is flowing. The resistivity of copper is $1.74 \times 10^{-8}\,\Omega$ m.

4.   Calculate the value of R and the power dissipated in R in the circuit of fig. 10.6.

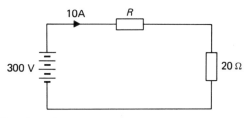

Fig. 10.6   for Problem 4

5.   Determine the current that flows when the following are switched on:
(a) a 6 W, 12 V soldering iron;
(b) a 60 W, 240 V light bulb;
(c) a 2 kW, 240 V kettle.

11 Chemical effects of electricity

The expected learning outcome of this chapter is that the student describes the chemical effects of electricity.

Introduction: examples of current being used for its chemical effect

The flow of electric current in certain materials causes a chemical effect. The most common example of the chemical effect of an electric current occurs in electrolytic solutions. An electrolytic solution, or electrolyte for short, is a solution of a substance, normally in water, which conducts electricity. Examples of electrolytes are solutions in water of acids, bases, and salts, such as sulphuric acid, caustic soda, copper sulphate, common salt. Note that not all solutions are conductors of electricity; for example, distilled water is a virtual non-conductor of electricity and is therefore not an electrolyte.

The chemical effect produced by a current flow through an electrolyte is called electrolysis. The actual chemical action is observed only at the terminals at which the current enters and leaves the electrolyte. These terminals are called electrodes and the electrode connected to the positive terminal (higher voltage) of the source is known as the anode, and the electrode connected to the negative terminal (lower voltage) of the source is known as the cathode.

A circuit which can be used to demonstrate the chemical effect of a current is shown in fig. 11.1(a). The electrolyte in this case is dilute sulphuric acid and the electrodes consist of platinum (platinum is used since it does not react chemically with sulphuric acid). When the switch S is closed, the ammeter reads current showing that there exists a continuous circuit and that conduction of current is occurring through the electrolyte. Bubbles of gas begin to appear at the electrodes and these gases may be collected in test tubes, as shown in the diagram. In fact, it can be ascertained that the gas at the anode is oxygen and the gas at the cathode is hydrogen. Oxygen and hydrogen are the constituent elements of water. Thus we can conclude that the electric current has produced a chemical effect by decomposing water into its constituent elements.

Another example of the chemical action of an electric current occurs when a current is passed through a solution of copper sulphate, *see* fig. 11.1(b). In this case gas is liberated at the anode (positive electrode) and pure copper is deposited on the cathode. This is an example of an important industrial process known as electroplating. In this process, the material which is to be plated (i.e. covered by a metal such as copper, nickel, chromium, silver) is made the

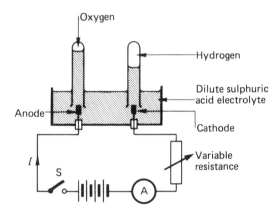

(a) Chemical effect of an electric current illustrated by gases appearing at cathode and anode when *I* flows

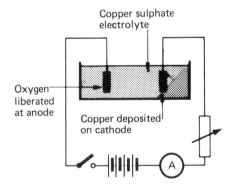

(b) Chemical effect of current in a copper sulphate electrolyte. Copper is deposited on cathode. If anode is platinum oxygen is liberated. If anode were copper, then copper is eaten away at the same rate as it is deposited on cathode

Fig. 11.1

cathode in an electrolyte which normally consists of a salt of the plating metal.

The anode can be made from the plating material so as to put back into the electrolyte the metal deposited on the cathode. The actual amount of metal deposited on the cathode, for a given metal plating material, depends on the current value and the time for which the current flows. These effects are used to calculate the thickness of plating material required.

The chemical effect of an electric current is made use of in:

electroplating
copper refining and production of aluminium
accumulators, e.g. car batteries and rechargeable batteries.

In a car battery (an example of a secondary cell to be discussed further in Section 11.5) there is a conversion of chemical energy to electrical energy and vice versa. When we charge a car battery or rechargeable batteries, such as those used in electronic calculators, burglar alarm systems, emergency lighting supply batteries, etc., we pass current into the positive terminal of the battery and force current to flow in the opposite direction to the current that the battery would deliver when it acts as a voltage source. Obviously, our charging source must have a greater e.m.f. than the battery voltage which is opposing the charging process. The electrical energy of our charger current is converted to chemical energy and thus we can identify an important application of the chemical effect of a current in converting electrical energy to chemical energy. When the battery is used to supply electricity, the chemical energy is reconverted to electrical energy.

11.1

The recognition of which materials are good conductors and which are bad conductors of electricity

The conduction of electricity through solid materials is by means of a flow of negatively charged electrons, as explained in Section 9.2. Metals are excellent conductors of electricity. Carbon in the form of graphite is also a good conductor of electricity. The ability of a material to conduct electricity is quantified by its resistivity (the resistance between the faces of a 1-metre cube of material, *see* Sections 9.17 and 9.18), thus the lower the resistivity of a material the greater its electrical conductivity. The resistivity of copper is $1.7 \times 10^{-8}\,\Omega\,m$, of graphite between 350 and $7000 \times 10^{-8}\,\Omega\,m$, of silicon (a semi-conductor material used for making transistors) between 10 and $10^4\,\Omega\,m$, and for porcelain (an insulator) about $10^{11}\,\Omega\,m$. So we can see that there is a tremendous range in the ability of solids to conduct electricity.

In the solid state the atoms of a substance are closely packed together and in the case of many materials the atoms are built up in regular crystalline patterns. When atoms form crystals in metals and in graphite the outermost electrons of the atoms are normally weakly bound and may break away from the atoms. These electrons are known as free electrons, since they are free to wander through the atomic lattice. When acted upon by an electrical force, for example by applying a battery across the material, a uniform drift is superimposed on their previously random paths. This drift of charged particles constitutes the electric current. As a further note, really beyond the specific learning objective of this section, we say in modern theory terms that these free electrons have available to them very closely separated energy states. They can therefore move from one energy state to another very easily when an electrical force is applied. In an insulating material there are far fewer free electrons but even so the energy states available to these electrons are very widely spaced. It thus requires an immense force to move an electron from one energy level to another. Hence such a material acts as a very poor conductor under normal values of applied voltage. It may however conduct electricity, technically known as 'breakdown', when several hundreds of kilovolts are applied.

11.2

The conduction of electricity in liquids is due to ions

The conduction in liquids is due to ions. Ions are charged atoms (atoms which have gained or lost one or more electrons) or groups of charged atoms (charged parts of a molecule). If no ions are present in a liquid then the liquid cannot conduct electricity. For example pure distilled water contains very few ions and is therefore a poor conductor of electricity (its resistivity is between 10^2 and $10^6 \, \Omega m$). Oil contains even fewer ions and is in fact used as an insulator, e.g. silicone oil has a resistivity of $10^{12} \, \Omega m$.

On the other hand liquid solutions of acids, salts, and bases in water readily conduct electricity. All these solutions contain ions which are the charge carriers of electrical current in the liquid.

For example, the crystals of salt are built up of ions carrying positive and negative charges. They are held in a crystalline array by attraction between these charges. When a salt is dissolved in water the attraction is reduced and the ions dissociate and take up a free existence in the solution. Let us take a specific example and consider the mechanism of current flow between two platinum plates placed in a solution of zinc

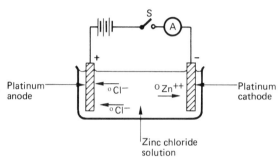

Fig. 11.2   Conduction of current in a solution of zinc chloride

chloride salt, as shown in fig. 11.2. We can state the following:

(1) Present in the solution are Zn^{++} ions (zinc atoms having lost two electrons) and Cl^- ions (chlorine atoms having gained one electron).
(2) When switch S is closed current flows in the circuit. In the solution Zn^{++} ions are attracted to the cathode (the platinum plate connected to the negative terminal of the battery source) and Cl^- ions are attracted to the anode (platinum plate connected to the positive terminal of the battery).
(3) When a Zn^{++} ion reaches the cathode it receives two electrons and is converted into a zinc atom. Thus zinc is deposited on the cathode.
(4) When a Cl^- ion reaches the anode it gives up its electron and combines with another chlorine atom (simultaneously or previously neutralised at the anode) to form a molecule of chlorine. Chlorine gas is liberated at the anode.
(5) Thus we can say that the mechanism of current flow in the solution is due to zinc and chlorine ions.

When silver nitrate ($AgNO_3$) is dissolved in water, silver ions (Ag^+) and nitrate ions (NO_3^-) are formed by the dissociation of some of the silver nitrate molecules. In the silver-plating process (see fig. 11.3(b) of next section) the silver ions are attracted to the cathode where they are neutralised and deposited.

An electron is taken from the battery for each silver ion deposited, and this electron neutralises a silver ion to form a silver atom. Nitrate ions are attracted to the anode where they each give up an electron and react with the silver anode to produce silver nitrate, thus replenishing the silver taken out of the solution by deposition at the cathode.

11.3
The principle of electrodeposition of metals: electroplating

Electroplating is the process whereby a layer of metal is deposited on an object by means of

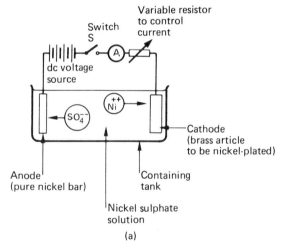

(a)

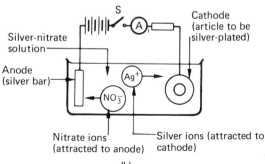

(b)

Fig. 11.3   Diagrams showing the basic electroplating process

electrolysis. The object to be plated is made the cathode, i.e. connected to the negative terminal of a battery or other d.c. voltage source. The anode (electrode connected to the positive terminal of the voltage source) is made of the plating metal. The electrolyte consists of a compound of the plating metal, a salt, dissolved in water. The electrolyte solution conducts electricity between the anode and cathode by means of positively charged metal plate ions and negatively charged ions present in the solution.

Two examples of electroplating: nickel-plating and silver-plating, may be described by reference to fig. 11.3. When switch S is closed current flows in the circuit and fine layers of metal are deposited from the electrolyte on the object (the cathode). At the same time an equal amount of metal is eaten away from the anode thus putting back into the electrolyte the metal lost by deposition on the cathode. By controlling the current and the time for which the current flows a required thickness of plating can be obtained. To

achieve a hard tenacious deposit the current density is normally limited to below $250 \, A/m^2$.

11.4
The description of a simple cell

A simple cell stores chemical energy which may be converted to electrical energy. The cell acts as a voltage source and may be used to drive electric current around an external circuit.

A cell in its simplest form is made by placing two plates or rods (the electrodes) of different materials in an electrolyte. The electrode materials are normally metals or a metal and carbon. The electrolyte is normally a solution of an acid, base, or salt, in water. Such cells are known as 'wet' cells. In a 'dry' cell the electrolyte is a water-moist paste rather than a liquid solution. When the electrodes are in the electrolyte a chemical reaction occurs between the individual electrodes and electrolyte. The result

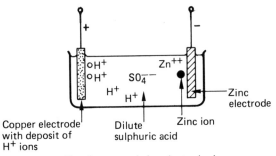

Copper electrode with deposit of H^+ ions

Dilute sulphuric acid

Zinc ion

Zinc electrode

(i) Copper and zinc electrodes in sulphuric acid to form a simple cell

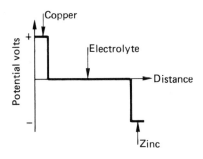

(ii) Potential-distance diagram for simple Cu-Zn cell on open-circuit

(a)

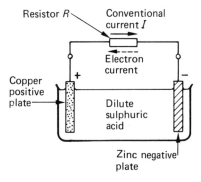

(i) Simple cell with a resistor R connected across its terminals

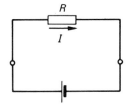

(ii) Equivalent circuit of above

(b)

Fig. 11.4 Simple cell diagrams

of these two reactions (one at each electrode) is that each electrode either gains or loses charge and hence a potential difference, known as an electrode potential, is established between each electrode and the electrolyte. The difference between the electrode potentials is equal to the e.m.f. of the cell. The e.m.f. of a cell depends on the nature of the electrodes and electrolyte used, on the concentration of the electrolyte, and on temperature.

Let us consider a practical example of a simple cell consisting of zinc and copper plates inserted into a solution of dilute sulphuric acid. H^+ ions from the sulphuric acid accumulate on the copper electrode making the copper plate positively charged. At the zinc electrode Zn^{++} ions enter the solution leaving the zinc plate negatively charged. Thus we have the state of affairs illustrated in fig. 11.4(a), with copper at a higher voltage with respect to the electrolyte and zinc at a lower voltage. The difference between these two voltages, that is the difference between their electrode potentials, is equal to the e.m.f. or open-circuit voltage of the cell. For example, if with a given concentration of sulphuric acid the electrode potential of the copper electrode is 0.5 V and that of the zinc is -0.5 V, the e.m.f. of the cell $= 0.5 - (-0.5) = 1.0$ V.

When a resistor is connected across the two electrodes, as shown in (b), the electrons left behind on the zinc plate by the zinc ions entering the solution can now flow through the resistor to the copper plate and neutralise the positively charged hydrogen ions accumulated there. For every two H^+ ions neutralised, one Zn^{++} ion enters into solution from the zinc electrode. Each of these Zn^{++} ions displaces in turn two more H^+ ions from the solution on to the copper plate. Thus an electron current is established between the zinc electrode via the resistor to the copper electrode. Since we regard by convention the direction of current flow as the direction in which positive charge moves, that is opposite in direction to the flow of electrons, we can say that the electron current is equivalent to a conventional current flow from copper via the resistor to zinc. The copper electrode acts as the positive and the zinc electrode as the negative terminal of the cell when the cell acts as a voltage source.

Finally two important practical points. The action at the zinc electrode eats away zinc and hence this electrode will eventually be 'consumed'. Hydrogen gas, formed by neutralisation of hydrogen ions, forms a layer at the copper plate which greatly increases the internal resistance of the cell and severely reduces the ability of the cell to provide current in an external circuit.

11.5
The difference between primary and secondary cells

A primary cell is one which uses up its active material during the time that it supplies current to a circuit. Eventually, most of the active material will be used up and then the cell can no longer give a useful supply of voltage. At this stage either the active material must be renewed or the cell must be thrown away. The dry battery used in torches, portable radios, etc. is an example of a primary cell.

A secondary cell is one in which the chemical action is reversible. When a cell supplies current, active material is used up, just as in the case of a primary cell. However, the active material may be subsequently restored by passing a current through the cell in the opposite direction to that of the current normally supplied by the cell. This process, known as charging, reverses the previous chemical process which took place when the cell was supplying current. The most common example of a secondary cell is the car battery. When we start a car, we draw a considerable amount of current from the battery. However, as soon as the car is in motion, an electromechanical generator (e.g. a dynamo) charges the battery, thus reversing the previous chemical action and restoring the active material in the battery.

A simple lead–acid secondary cell can be constructed by inserting two lead plates into a solution of dilute sulphuric acid. A diagram of the basic cell is shown in fig. 11.5(a).

Initially, zero voltage exists between plates. The cell must first be charged using an external voltage source. Figure 11.5(b) gives a diagram of a simple charging circuit. The plate connected to the positive terminal of the external source will, when the cell is charged, be the positive terminal of the secondary cell, and the other plate connected to the lower voltage terminal of the source will become the negative terminal.

During the charging process, the positive plate (the anode) is coated with a film of active material and this material is responsible for establishing an e.m.f. in the cell. The charging current in the

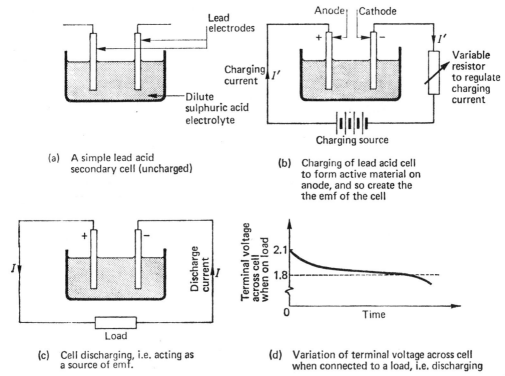

(a) A simple lead acid secondary cell (uncharged)

(b) Charging of lead acid cell to form active material on anode, and so create the the emf of the cell

(c) Cell discharging, i.e. acting as a source of emf.

(d) Variation of terminal voltage across cell when connected to a load, i.e. discharging

Fig. 11.5

electrolyte is due to positive hydrogen and negative sulphate ions. The sulphate ions are attracted towards the positive plate, where they give up their charge. Oxygen, which is liberated when the ions give up their charge, then reacts with the lead and forms a brown coating of lead peroxide—the active material. The hydrogen ions are attracted to the negative plate (the cathode), where they give up their charge. Hydrogen gas is thus liberated at the cathode, but no chemical action occurs at this plate.

After this first charging process, the cell has an e.m.f. which is produced by the chemical effect of the charging current on the anode. The cell could now be used as a source to supply current to an external load, as shown in fig. 11.5(c), although a simple cell of this type, with just a single charge, would be unable to sustain a current for very long. When the cell supplies current, we say that the cell is undergoing discharge. However, if the process of charge followed by discharge is repeated, the capacity of the cell to supply current is steadily increased, since more and more lead at the positive plate is transformed into the active lead peroxide material. In practice, this method of continual charge and discharge is only employed in the production of

accumulator plates; an accumulator is the term used to describe a storage battery composed of secondary cells. A practical lead–acid accumulator would already have active material deposited on its anode and would require only a single charge before it could be used.

When the cell discharges, the lead peroxide on the positive plate is gradually changed into a coating of lead sulphate and water. Thus, the sulphuric acid electrolyte becomes diluted and hence its relative density falls. In fact, the state of charge or discharge of a cell is measured by noting the relative density of its sulphuric acid electrolyte. When fully charged, the relative density is about 1.26 and the cell e.m.f. is 2.1 V. Actually, the e.m.f. can rise to about 2.6 V on the final stage of charge, although on discharge the e.m.f. will fall very rapidly to 2.1 V. The e.m.f. of the cell falls on discharge, and in practice should not be allowed to drop below 1.8 V, when the relative density of the electrolyte is approximately 1.17. The discharge curve (terminal voltage versus time) of a lead-acid cell is shown in fig. 11.5(d). During discharge, lead sulphate is also formed on the cathode.

On a subsequent charge, current entering the positive plate from the electrolyte in the form of

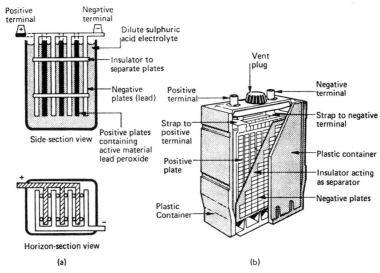

Fig. 11.6   Lead-acid secondary cells

sulphate ions attacks the lead sulphate deposit and changes it to the active lead peroxide material. At the same time, hydrogen ions react with the lead sulphate coating on the cathode, converting this back into lead and also liberating hydrogen gas. At both plates, sulphuric acid is produced, thus replenishing the strength of the electrolyte.

Although commercial lead–acid cells are capable of supplying several amperes (and even several tens of amperes for a short time) care must be taken not to exceed the rated current value for a cell. If this were exceeded, buckling of the cell plates might occur and as a consequence the cell would be permanently damaged.

The capacity of a lead–acid accumulator is stated in terms of the quantity of electricity it can supply when uniformly discharged in a given time, assuming, of course, that it is initially fully charged and is finally not discharged beyond the point that it could not be fully charged again. The capacity is normally quoted in ampere-hours (A h). Thus a 40 A h cell could supply 2 A for 20 hours or 4 A for 10 hours (provided of course that the discharge current does not exceed the current rating). Strictly speaking the ampere-hour is not a recognised SI unit, but its use to specify cell capacity is very widespread. Note that

$$1\,A\,h = 1\,A \times (60 \times 60)\,s$$

$$= 3600\,As\ (ampere\text{-}seconds)$$

$$= 3600\,C\ (coulombs)$$

A labelled diagram of a typical lead–acid accumulator is shown in fig. 11.6. The positive plate consists of the active lead peroxide material in the form of a paste pressed into a lead–antimony alloy (94 % lead, 6 % antimony) framework. The negative plate consists of a similar structure carrying lead in spongy form. To ensure a low internal resistance, a large plate surface area is needed, and to achieve this a number of positive and negative plates are interleaved, e.g. in fig. 11.6(a) there are four negative and three positive plates. The respective positive and negative plates are joined by a lead bar. To further reduce the internal resistance, the plates should be close together, but must be prevented from touching each other. Separators of plastic, glass, and 'treated' wood insulating materials, are used for this purpose. The electrolyte is dilute sulphuric acid which should have a relative density of about 1.21 when the cell is fully charged. The container itself is normally made of a hard durable plastic.

11.6
The variation of source potential difference of a cell with time and when different loads are used

(a) The 'no-load' voltage of e.m.f. of a source and the concept of internal resistance

The voltage measured across the two terminals of a source of electrical energy, for example, a cell, when these terminals are not connected to any external component or network, is equal to the

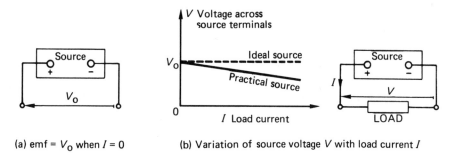

(a) emf = V_0 when $I = 0$ (b) Variation of source voltage V with load current I

Fig. 11.7

e.m.f. of the source. Thus, in fig. 11.7(a), the voltage measured across the open terminals (assuming that the voltmeter takes negligible current), V_0 = e.m.f. of the source. Under these conditions, no current flows and we say that the source is on no-load, or open-circuited. The term 'load' is used in electrical circuits to denote the component or network connected across the source terminals.

If our source is ideal, the voltage across its terminals remains constant and equal to the e.m.f. or no-load value of voltage when the source actually delivers current to a load connected across its terminals. However, for a practical source, the terminal voltage decreases as the load current increases. This effect is shown graphically in fig. 11.7(b).

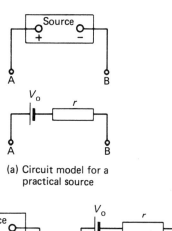

(a) Circuit model for a practical source

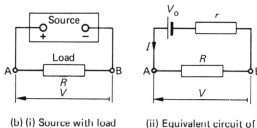

(b) (i) Source with load   (ii) Equivalent circuit of source with load

Fig. 11.8

Let us attempt to give an explanation of this decrease in terminal voltage with increasing load current by constructing a simple circuit model for a practical source, and then using this model to represent the electrical performance of the source in a circuit. A circuit model of a device must simulate accurately the behaviour of the device in an electrical circuit. In fact, one of the simplest examples of a circuit model is that of a linear resistor, defined by Ohm's Law as $R = V/I$, which simulates the behaviour of a conductor in a circuit.

A practical source will always exhibit some resistance to the flow of current and hence energy will be expended in driving current within the source itself as well as in the externally connected circuit components. This effect may be taken into account by representing the source between its two terminals by a model consisting of a voltage equal to the source e.m.f. of V_0 volts in series with a resistance r. The circuit model for a practical source is shown in fig. 11.8(a). The resistance r is known as the *internal resistance* of the source since it simulates the resistance or heat dissipation effects within the source.

Next, consider this source connected in series with a resistor R. The circuit diagram for this situation is shown in fig. 11.8(b). We can, of course, analyse this circuit straight away using Ohm's Law and the fact that the applied voltage equals the sum of the voltages across the individual resistors, i.e.

$$V_0 = RI + rI = (R + r)I$$

so

$$I = \frac{V_0}{R + r}$$

and the voltage across terminals A–B of the source,

$$V = RI = \frac{R}{R + r} V_0$$

Thus if the load R is much greater than the internal resistance r,

$$\frac{R}{R+r} \approx 1 \quad \text{and so the terminal voltage } V \approx V_0$$

(the symbol $\approx$ means approximately equal to). However, if $R = r$

$$\frac{R}{R+r} = \frac{R}{2R} = \tfrac{1}{2} \quad \text{so} \quad V = \tfrac{1}{2}V_0 \text{ (half the e.m.f.)},$$

whilst if $R = 0.1r$ (i.e. load very much smaller than the internal resistance), then

$$\frac{R}{R+r} = \frac{0.1r}{0.1r+r} = \frac{0.1r}{1.1r} = \tfrac{1}{11} \quad \text{and} \quad V = \tfrac{1}{11}V_0,$$

showing that the terminal voltage is less than 10% of the e.m.f. or no-load voltage.

Let us also consider the circuit from a power (or energy flow) point of view. One reason for doing this is that the formal definition of the e.m.f. of a source is often stated in terms of energy flow per coulomb of charge transported:

The e.m.f. of a source is equal to the energy supplied by the source to transport one coulomb of charge completely around the circuit to which the source is connected. Thus, using this definition, the total energy per second (i.e. the power) supplied by the source

$$= (\text{e.m.f.}) \times (\text{charge per second}) = V_0 I.$$

The power dissipated within the source (as heat $= rI^2$,
the power dissipated in the load resistor $= RI^2$,
and as energy and power are conserved,
power supplied by source $=$ power dissipated in r + power dissipated in R
so

$$V_0 I = rI^2 + RI^2.$$

Cancelling through the above equation by I gives

$$V_0 = rI + RI$$

from which the source terminal voltage

$$V = RI = V_0 - rI$$

The above results, obtained by analyses in which we represented a practical voltage source by an e.m.f. V_0 volts in series with an internal resistance r ohms, may be used to provide an explanation for the characteristics of a practical source. It predicts, correctly, that source potential difference falls with increasing load current.

Unfortunately, our simple model cannot explain all the phenomena associated with a practical source. It would predict that for a fixed load current I, the source potential difference V should also be fixed. However, V for most cells also tends to decrease with time even for a fixed value of a load resistor. The reason for this is that the chemical activity generated in the cell when supplying current tends to increase the effective value of the internal resistance r, thus reducing V still further, as illustrated in fig. 11.5(d).

The internal resistance of a 1.5-V dry battery is normally of the order of a few ohms, whilst the internal resistance of a freshly charged 2-V accumulator is of the order of 0.01 Ω. Some cells, for example a Weston cell which is used as a standard of voltage, have very high internal resistances of the order of several hundred ohms.

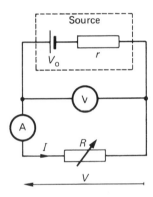

(a)   Circuit for measurement of load current versus terminal voltage graph

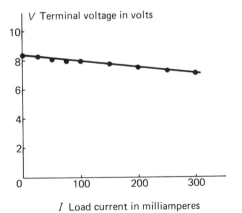

(b)   Graph of terminal voltage of a source (dry battery) versus load current

Fig. 11.9

(b) An investigation of the effect of load current on terminal voltage and the determination of the internal resistance of a source

The measurement of the effect of the load current I on the terminal voltage V of a source can be investigated by means of the circuit shown in fig. 11.9(a). A high resistance voltmeter is used to measure the terminal voltage and an ammeter is connected in series with the variable load R to measure the load current. Since the source e.m.f. equals the sum of the voltages across its internal resistance r and the load resistance R, we have

$$V_0 = RI + rI$$

so the terminal voltage, $V = RI = V_0 - rI$.

Since the e.m.f. V_0 can be measured on no-load (i.e. the voltage across source terminals with R disconnected) and V and I are measured in the circuit we can determine the internal resistance from the above relation, i.e.

$$V_0 - V = rI \quad \text{so} \quad r = \frac{V_0 - V}{I}$$

Some experimental results for measurements taken on an 8 V dry battery are given below:

terminal voltage V (volts)	8.2	8.1	8.0	7.9	7.8
load current I (milliamperes)	0	25	50	75	100

terminal voltage V (volts)	7.65	7.45	7.3	7.1
load current I (milliamperes)	150	200	250	300

The load current values were obtained by adjustment of the variable load R. Care was taken not to exceed the rated value of current for the battery, in this case a maximum of 300 mA.

The graph of the terminal voltage V versus load current I using the above results is plotted in fig. 11.9(b). The graph shows that V falls as I increases, showing the effect of internal resistance dropping voltage within the source. A determination of r may be found by substituting $V_0 = 8.2$ V and for example $V = 7.1$ V, $I = 0.3$ A (last pair of results in above table) in

$$r = \frac{V_0 - V}{I} = \frac{8.2 - 7.1}{0.3} = \frac{1.1}{0.3} = 3.7\,\Omega.$$

Example 11.1
The no-load voltage measured across the terminals of a source was 20.0 V. When a 50 Ω load was connected across the source, the terminal voltage was observed to fall to 18.0 V. Determine the e.m.f. and internal resistance of the source.

Solution
Let $V_0 = $ e.m.f. and $r = $ internal resistance of source.
The terminal voltage on no-load = source e.m.f., so $V_0 = 20$ V.
When $R = 50\,\Omega$ load is connected across source, terminal voltage $V = 18$ V, so load current

$$I = \frac{V}{R} = \frac{18}{50} = 0.36\,\text{A}.$$

The remainder of the e.m.f., i.e. $20 - 18 = 2$ V, is dropped across the internal resistance, so

$$rI = 2 \quad \text{and} \quad r = \frac{2}{I} = \frac{2}{0.36} = 5.6\,\Omega.$$

Check: using the formula

$$r = \frac{V_0 - V}{I} \quad \text{where} \quad I = \frac{V}{R} = \frac{18}{50} = 0.36\,\text{A},$$

we obtain

$$r = \frac{20 - 18}{0.36} = \frac{2}{0.36} = 5.6\,\Omega.$$

Problems 11: Chemical effects of electricity

1. Give an explanation of the fact that copper is an excellent conductor of electricity whereas porcelain is a virtual non-conductor of electricity.
2. Describe the mechanism of conduction of electricity in an electrolyte. Explain why distilled water is a poor, whilst a solution of salt in water is a relatively good, conductor of electricity.
3. Describe, with the aid of a suitable diagram, the process of electroplating.
4. Distinguish between a primary and a secondary cell.
   Draw a labelled diagram of a lead–acid cell accumulator and describe how it may be charged. Sketch also a typical curve showing the variation of terminal voltage with time when (i) the accumulator is discharging, (ii) the accumulator is charging.
5. Describe an experiment to demonstrate that

the terminal voltage of a source falls with increasing load current.

In such an experiment the following measurements were recorded:

Load current (A)	0	0.5	1.0	1.5	2.0	2.5
Terminal voltage (V)	10	9.7	9.1	8.8	8.3	8.1

Plot the graph of terminal voltage versus load current, drawing the best straight line through the plotted points. Determine from the graph the e.m.f. and the internal resistance of the source used in the experiment.

6.   Define the internal resistance and e.m.f. of a practical voltage source. A battery is connected in series with a variable resistor and an ammeter. The following measurements were noted:

Ammeter reading (mA)	75	50	25	15
Resistor setting (Ω)	18	28	58	98

Draw the circuit diagram using a suitable circuit model to represent the battery and use the table of results to determine the e.m.f. and internal resistance of the battery.

12 Magnetic and electromagnetic effects

The expected learning outcome of this chapter is that the student describes magnetic and electro-magnetic effects and simple applications.

12.1
A current-carrying conductor produces a magnetic field and examples of applications of this effect

In 1820 Oersted found by experiment that the needle of a magnetic compass was deflected when brought close to a conductor carrying current. He thus established that an electric current produces a magnetic effect.

A simple illustration of the magnetic effect of a current is shown in fig. 12.1. When the switch S is open, no current flows in the circuit and the compass needle points in the direction N of magnetic north. When the switch S is closed, the circuit is complete, current flows in the conductor and the compass needle is deflected to a new

position. If the direction of the current is reversed, as in fig. 12.1(b), then the direction of the deflection of the compass needle is also reversed.

The magnetic effect of an electric current has many extremely important applications. Some examples of its uses are stated below.

(a) Electromagnets

If a number of turns of conducting wire are wound around an iron bar (or a bar of other magnetic material), and a current is established through the wire, the bar behaves as a magnet. The strength of this magnet increases with the strength of the current. When the current is switched off, the bar loses its magnetic proper-ties. This principle is used in the electromagnet which consists, in its simplest form, of a soft-iron bar (often the term core is used to describe this) on which is wound a coil of insulated wire, as shown in fig. 12.2. When current flows in the coil, the electromagnet is capable of attracting and lifting pieces of iron and other magnetic materials. When the current is switched off, the electromagnet ceases to possess its magnetic property and will release any material that it had previously attracted. Hence one obvious appli-cation of an electromagnet is in scrap-yards to move iron and other magnetic materials.

(b) Magnetic relays

Figure 12.3(a) shows an example of the magnetic effect of a current being used in a relay switching application. When a relatively small current I flows through the relay windings, the soft-iron core P is attracted downwards, thus closing the break in the main circuit by placing a conducting bar between terminals A and B. This bar completes the main circuit and hence allows the main circuit current to flow. On opening switch S in the relay circuit, I falls to zero with the result that its magnetic effect in the windings ceases to exist. The return spring forces the conductor bar upwards, breaking and switching off the main circuit.

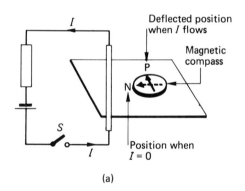

(a)

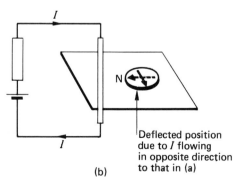

(b)

Fig. 12.1

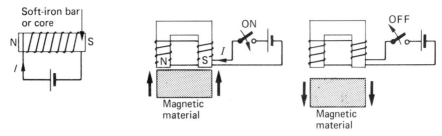

(a) Simple electromagnet   (b) Electromagnet for 'crane' working

Fig. 12.2

In the electrical bell circuit of fig. 12.3(b), the iron disc attached to the bell striker is attracted towards the windings whenever current flows. However, this process breaks the circuit, the magnetic effect of the windings ceases to exist and hence the spring returns the striker to its original position. The cycle of events is repeated with the result of ringing the bell at each forward stroke.

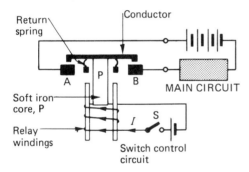

(a)   Relay working on magnetic effect of a current

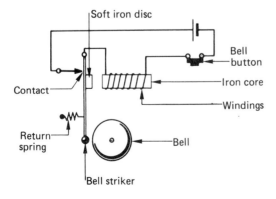

(b) Electric bell circuit

Fig. 12.3

(c) *Moving coil and soft-iron ammeters and voltmeters*

The magnetic effect of a current in a coil depends on the strength of the current, and a current-carrying conductor experiences a force when placed close to a magnet. These properties are used to measure current. The basic operation of a moving coil ammeter is explained in Section 12.4.

The magnetic effect of an electric current is also made use of in: electric motors (and, indirectly, generators and transformers); magnetic tapes (for sound recording and computer storage use); loudspeakers (and, indirectly, microphones and pick-ups); and magnetic focusing and deflection coils (e.g. as used to deflect an electron beam in a television receiver).

We identify that the magnetic effect of a current is employed in the above applications since in every case current is used in some way to increase, vary, or interact with the magnetism or magnetic properties associated with the constituent components of the device. Sometimes this is fairly simple to see, as in the electromagnet. Sometimes the effect is more complex as, for example, in an electric motor, an ammeter, or a loudspeaker. In the last case, a varying electrical current in a coil attached to a loudspeaker cone interacts with the magnetism produced by a permanent magnet placed round the coil and this interaction causes the loudspeaker cone to vibrate and produce sound.

12.2
The description of the type of a magnetic field produced by (a) a bar magnet, and (b) a solenoid

The magnetic field of a simple bar magnet is very similar to that produced by a solenoidal coil when the coil is carrying direct current.

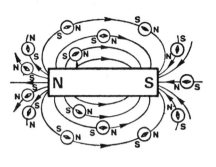

(a) Magnetic field pattern around a bar magnet

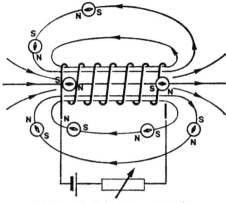

(b) Magnetic field pattern around a solenoid

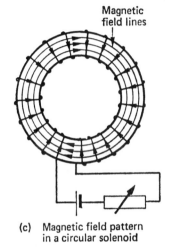

Magnetic field lines

(c) Magnetic field pattern in a circular solenoid

(d) "Rule" for finding S - N pole of a solenoid: when current viewed from one end circulates clockwise that end acts as a S pole. If current circulates anticlockwise then the end acts as a N pole.

Fig. 12.4

The forms of these fields may be investigated by plotting the deflection direction of a small magnetic compass needle at various points in the space surrounding the bar magnet and solenoid. The magnitude and direction of the magnetic field experienced by the test compass needle will vary with position. The density or 'closeness' of these lines gives a rough measure of field strengths; the closer the lines, the greater the field strength. Although we have no means of actually measuring the magnetic field strength with our test needle we can plot the lines of magnetic flux as follows.

Place the test compass needle close to the north pole of the bar magnet and record the direction in which the needle points by a small arrow. Repeat for a series of other points at reasonably close intervals and join up—the tip of one to the base of the next—to form a continuous contour or path which it will be observed goes from one pole of the magnet to the other. This contour is known as a magnetic field, or flux, line. By repeating the procedure, other field lines may be drawn in and a complete set surrounding the magnet obtained. The set of these magnetic field lines is known as the magnetic field pattern. The magnetic field pattern of a bar magnet, together with those for solenoids, are shown in fig. 12.4. The direction of the field or flux lines is taken as the direction in which the north pole of the magnet compass would point.

The magnetic field patterns due to a solenoid may be obtained in a similar manner to that described above. In fact, the field lines could be plotted within the solenoid itself, showing that they form closed loops. The two ends of a straight solenoid act as magnetic poles. A rule for finding the respective north and south poles of

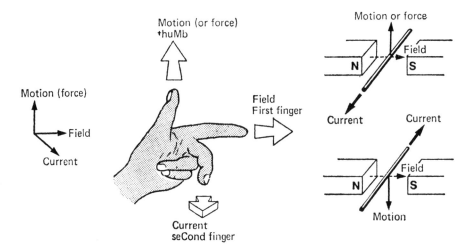

(a) Fleming's left hand rule to determine direction of motion   (b) Examples showing the
     of a current carrying conductor in a magnetic field   direction of the force on
                                                                       a conductor

Fig. 12.5

the ends with respect to the direction of current flow is given in fig. 12.4(d).

12.3
A current-carrying conductor experiences a force in a magnetic field

A current-carrying conductor experiences a force in a magnetic field. This force is proportional to the current and length of the conductor and increases with the strength of the magnetic field.

The direction of the force is at right angles to both the directions of the current and magnetic field. A useful aid for determining the direction of the force on a current-carrying conductor is given by Fleming's left-hand rule, illustrated in fig. 12.5(a): if the first finger of the left hand is pointed along the direction of the magnetic field lines and the second finger in the direction of current flow, then the thumb gives the direction of the force on the conductor (or if the conductor were free to move, the direction of its motion). The rule is applied in fig. 12.5(b). Check to see that you agree with the direction of the forces shown.

A simple experimental set-up to demonstrate the forces on a current-carrying conductor in a magnetic field is shown in fig. 12.6. The conductor consists of a fine length of wire ('fine' to enable the effect of the force on the conductor to be more readily seen) clamped between two fixed

supports. The magnetic field is produced by either (a) a bar magnet, or (b) a solenoid.

The following results may be obtained:

(i) When the current flows in the wire, it moves downwards in the figures shown. If the bar magnet is turned through 180° or the direction of the current in the solenoid is reversed (i.e. the magnetic field direction is reversed) the wire moves upwards.
(ii) The direction of movement of the conductor is also reversed if the direction of current in the conductor is reversed.
(iii) The displacement of the conductor in the vertical plane is increased when the current is increased and when the magnetic field strength is increased. The former may be demonstrated by varying the current by means of the variable resistor R, the latter by bringing the bar magnet or solenoid closer to the wire.

12.4
Basic operation of a moving-coil meter and a simple d.c. motor

(a) *The explanation of the basic operation of a moving-coil meter*

The force on a conductor in a given magnetic field increases linearly with the current passing through it, e.g. if the current is doubled, the force

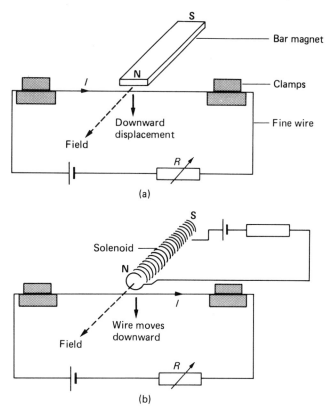

Fig. 12.6

(a)

(b)

doubles. This property provides a very convenient means of measuring current (and voltage) and is applied practically in moving-coil meters.

The essential features of a moving-coil meter are shown in fig. 12.7. The small coil through which the current to be measured flows is wound on an aluminium former, which is free to rotate against the return spring between a soft-iron core and pole pieces N and S of a permanent magnet. The magnet (with the help of the core) provides a radial field which is always perpendicular to the coil.

When a current flows through the coil, the conductor sections, ab and cd (see fig. 12.7(c)), experience a force. The force on ab is upwards, the force on cd is downwards. There are no forces on sections bd and ac. The forces on ab and cd will cause the coil to move against the restoring forces of the spring. In equilibrium, the twisting effect of these forces will equal the restoring effect of the spring, and the coil rotation measured by a pointer attached to the coil shaft will give a measure of the current flowing. In practice, a moving-coil meter would be calibrated using known values of currents.

(b) The explanation of the basic operation of a simple d.c. motor

The force on a current-carrying conductor in a magnetic field forms the basic factor whereby electrical energy can be converted to mechanical energy in electric motors.

Figure 12.8(a) shows a diagram of a simple d.c. motor. This motor consists of a coil of one turn situated between the poles of a powerful magnet. Attached to the coil is a shaft and the coil plus shaft are free to rotate. Current is supplied to the coil from a d.c. voltage source by means of a commutator (see fig. 12.8(b)), the two segments of which are insulated from each other but permanently connected to the opposite ends of the coil.

When current is fed into the coil in the direction shown in fig. 12.8(c)(i) we find on applying Fleming's left-hand rule that conductor wx experiences an upward force and conductor yz a downward force. This causes the coil to rotate. When the coil reaches the vertical position, the current in these two conductors is reversed by means of the commutator and hence the forces are all also reversed in direction. Thus,

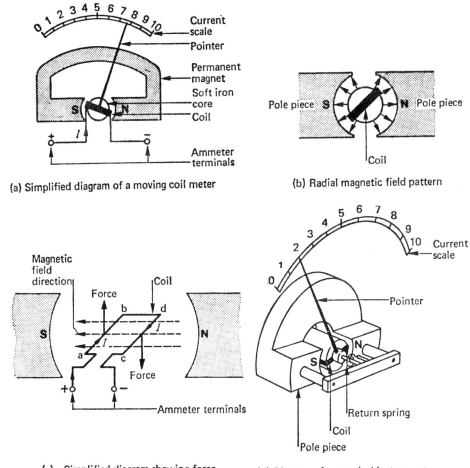

(a) Simplified diagram of a moving coil meter

(b) Radial magnetic field pattern

(c) Simplified diagram showing force on coil

(d) Diagram of a practical instrument

Fig. 12.7

the coil continues to rotate. Actually, when the coil is in the vertical position, the forces on the coil conductors wx and yz do not cause rotation. However, the rotational energy carries the coil over the vertical position and as the current is also reversed at this instant, rotation is maintained.

12.5
The description of electromagnetic induction with reference to the movement of a magnet in a coil connected to a meter

Let us first state what is meant by electromagnetic induction. Electromagnetic induction is the production of an e.m.f. or voltage in a circuit or conductor by a change in the magnetic flux 'linking' the circuit, or by a change of the magnetic flux 'cutting' a conductor. Figure 12.9 illustrates the generation of an e.m.f. by the movement of a magnet in and out of a coil. The movement of the magnet in this case causes a change in the flux linking the coil and thus induces an e.m.f. in the coil.

When the magnet M in fig. 12.9(a) is moved towards the coil of wire C, the galvanometer G (a sensitive ammeter which in this case has a centre-zero scale) connected across the terminals of the coil gives a kick and registers a deflection. The deflection falls to zero as soon as the magnet comes to rest. If M is withdrawn from the coil, the galvanometer registers a deflection in the opposite direction. Although it may be difficult to observe experimentally, the magnitude of the galvanometer deflections depends on the speed of moving the magnet—the faster, the greater the deflection. By moving the magnet continuously

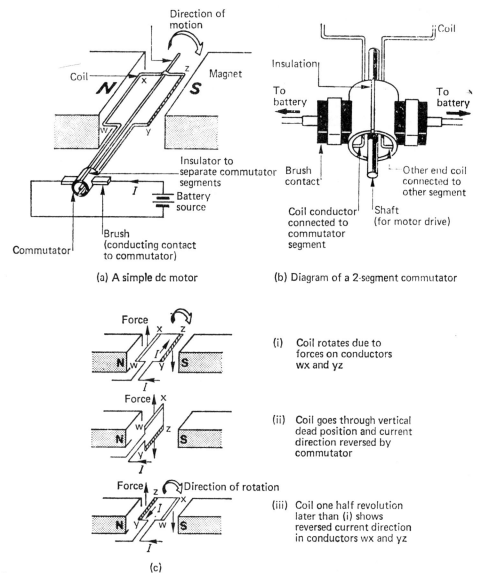

Fig. 12.8

(a) A simple dc motor

(b) Diagram of a 2-segment commutator

(i) Coil rotates due to forces on conductors wx and yz

(ii) Coil goes through vertical dead position and current direction reversed by commutator

(iii) Coil one half revolution later than (i) shows reversed current direction in conductors wx and yz

(c)

in and out of the coil, the galvanometer needle can be made to oscillate backwards and forwards about its zero position and a relatively large swing in needle deflection can be obtained. In fact, an identical experiment to this was undertaken in 1831 by Faraday in his classic work on electromagnetic induction.

From the observations of our experiment, we can make the following conclusions:

(1) The movement of a magnet in a coil generates an e.m.f. and thus causes current to flow in the coil.

(2) The direction of action of the e.m.f., and therefore the current induced in the coil, depends on the direction of motion of the magnet.

These observations form the basis of Faraday's Law of electromagnetic induction, which states:

The e.m.f. induced in a circuit is equal to the rate of change of magnetic flux (magnetic field lines) through the circuit; or the rate at which conductors in the circuit 'cut' the magnetic flux.

Although the formal definition of magnetic flux is not a specific learning objective in this level 1

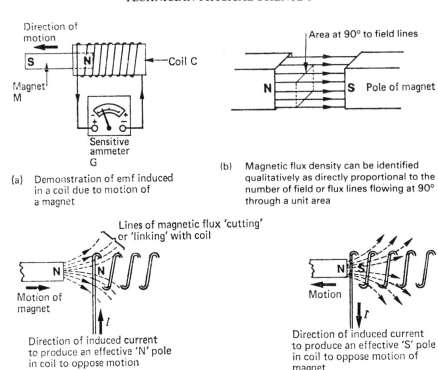

(a)   Demonstration of emf induced
in a coil due to motion of
a magnet

(b)   Magnetic flux density can be identified
qualitatively as directly proportional to the
number of field or flux lines flowing at 90°
through a unit area

Lines of magnetic flux 'cutting'
or 'linking' with coil

Motion of
magnet

Direction of induced current
to produce an effective 'N' pole
in coil to oppose motion
of magnet

Motion

Direction of induced current
to produce an effective 'S' pole
in coil to oppose motion of
magnet

(c)   Magnetic flux cutting or linking with a coil
and the direction of the induced current
due to the changing flux produced
by the motion of the magnet

Fig. 12.9

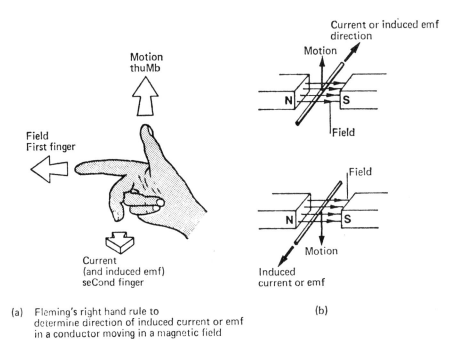

(a)   Fleming's right hand rule to
determine direction of induced current or emf
in a conductor moving in a magnetic field

(b)

Fig. 12.10

unit, we can identify magnetic flux as being directly proportional to the number of magnetic field or flux lines flowing at right angles through a unit area in the magnetic field (*see* fig. 12.9(b)). Thus, in the case of a magnet moving within a coil, the coil experiences a changing magnetic flux due to the magnet's motion, and the rate of change of flux 'linking' the coil produces an e.m.f. in the coil. If the magnet is held stationary and the coil is moved in its magnetic field, an e.m.f. is also induced in the coil. In both cases the coil experiences a changing magnetic flux and it is this that induces the e.m.f.

The direction of the induced e.m.f. is given by Lenz's Law, which states:

The direction of the induced e.m.f. is such as to cause a current, if the circuit is complete,

in such a direction so as to set up magnetic flux in the opposite direction to the magnetic flux producing the e.m.f.; or equivalently, to set up a magnetic field in the circuit to oppose the motion of the source of magnetic field inducing the e.m.f.

In the case of a magnet approaching a coil, an e.m.f. and therefore current will be induced in the coil so as to produce an effective like pole at the coil end closest to the approaching magnet and thus oppose the magnet's motion. When the magnet is withdrawn, the e.m.f. and current direction will be reversed to set up an unlike pole and thus continue to oppose the magnet's motion. These effects are illustrated in fig. 12.9(c).

When a conductor moves through a magnetic field, it is said to be cutting magnetic flux. In such

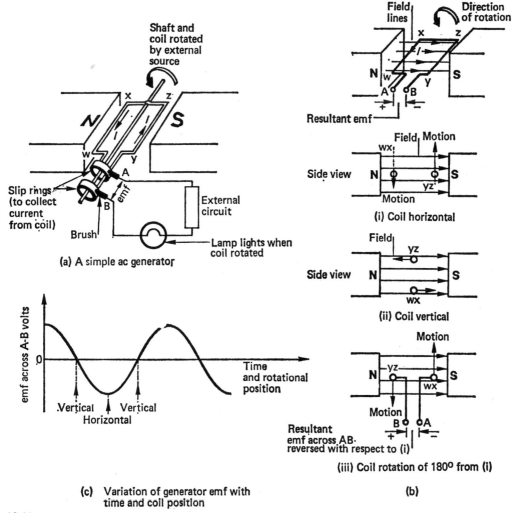

(a) A simple ac generator

(c) Variation of generator emf with time and coil position

(i) Coil horizontal

(ii) Coil vertical

(iii) Coil rotation of 180° from (i)

(b)

Fig. 12.11

cases an e.m.f. is also induced in the conductor, and current will flow if the circuit is complete. The direction of this induced e.m.f. may be determined from Fleming's right-hand rule (*see* fig. 12.10(a)): hold your right hand with thumb and first two fingers at right angles; point the first finger in the direction of the magnetic field lines, the thumb in the direction of motion, then the second finger will point in the direction of the induced e.m.f. and current. Check, by applying the rule, that you agree with the direction of induced e.m.f. and current shown in the two examples of fig. 12.10(b).

12.6
The explanation of the basic operation of an a.c. generator

Figure 12.11(a) shows a diagram of a simple a.c. generator in which an alternating e.m.f. is generated by rotating a coil in the magnetic field of a powerful magnet. In an electric generator external mechanical energy is supplied and converted into electrical energy.

The basic operation of the simple generator may be explained as follows. Conductors wx and yz cut the magnet flux produced by the magnet as they are rotated and thus an e.m.f. is induced in the conductors. The directions of the induced e.m.f. and current (found by applying Fleming's right-hand rule) for three consecutive positions in the rotation cycle are shown in fig. 12.11(b):

(1) in this position the e.m.f. generated in both conductors is at a maximum since at this instant they are cutting flux at the maximum rate (remember, e.m.f. = rate at which conductor cuts flux),

(2) when the coil is vertical, zero e.m.f. is generated since both conductors are moving at this instant parallel to the lines of magnetic flux and therefore not cutting flux,

(3) on moving through the vertical the e.m.f.s induced in wx and yz are reversed and again maximum in value when the conductors are in the horizontal position.

The output e.m.f. of the generator is equal to the sum of the e.m.f.s induced in wx and yz. It is developed across the ends of the coil which are connected to copper slip rings. Conducting brushes (e.g. carbon) A and B pressing on these slip rings, and in permanent contact with the rings as these rotate with the coil, connect the e.m.f. to an external circuit or load.

The variation of generated e.m.f. with coil position and time is plotted in fig. 12.11(c). The variation is in the form of a cosine wave with maximum voltage values when the coil is horizontal and zero when the coil is vertical and changing sign when the coil goes through the vertical position.

Problems 12:   Magnetic and electromagnetic effects

1.   State the three main effects of an electric current and identify which of the effects is being made use of in each of the following: charging a battery; fuse wire; magnetic tape; loudspeaker; an electric soldering iron. Give two further applications for each of the three main effects.

2.   (a) Sketch the magnetic field produced by (i) a bar magnet, (ii) a straight solenoid.

(b) Describe an experiment to demonstrate that a current-carrying conductor experiences a force in a magnetic field.

3.   Draw a labelled diagram of (a) a moving coil ammeter, (b) a simple d.c. motor. Explain their basic mode of operation.

4.   Describe an experiment to demonstrate what happens when a permanent magnet is moved in a coil of wire connected to a galvanometer.

5.   Draw a labelled diagram of a simple a.c. generator. Explain its basic mode of operation and sketch a graph of the e.m.f. produced by the generator against time.

Answers to selected problems

Problems 2
2 (a) $25 \text{ N/mm}^2 = 25 \text{ MN/m}^2$
   (b) 8 mm
3 (a) 0.45 mm   (b) 0.1 mm
   (c) 66.7 N; 1.4 mm
4 50 MN/m^2 or $50 \times 10^6 \text{ N/m}^2$, 286 mm^2
5 (a) 0.42 mm   (b) 964.3 MN/m^2 or
   $964.3 \times 10^6 \text{ N/m}^2$

Problems 3
5 (a) 245.3 N   (b) 4.125 m from P
6 1.6 kg
7 392 Nm, 147 Nm; clockwise rotation;
   8.33 kg

Problems 4
1 (a) $132.4 \times 10^3 \text{ N}$; $22.1 \times 10^3 \text{ N/m}^2$
   (b) $11.5 \times 10^3 \text{ N/m}^2$
2 (a) false   (b) false   (c) true
   (d) false   (e) true
3 $(15.08 + 1)10^5 \text{ Pa} = 16.08 \times 10^5 \text{ Pa } (\text{N/m}^2)$
4 (a) $16 \times 10^3 \text{ N/m}^2$ above atmospheric
   pressure, i.e. $(16 + 101.3)10^3 = 117.3 \times$
   10^3 N/m^2;
   (b) $5.3 \times 10^3 \text{ N/m}^2$ below atmospheric
   pressure, i.e. $(101.3 - 5.3)10^3 = 96 \times$
   10^3 N/m^2;
   (c) at atmospheric pressure,
   $101.3 \times 10^3 \text{ N/m}^2$

Problems 5
1 (a) 22.2 m/s   (b) 5.88 m/s
   (c) 6.04 s
2 (b) 0.5 m/s; 0.2 m/s   (c) 1.2 m/s
   (d) 50 s; 0 s
3 (a) 3 m/s^2 (b) -5 m/s^2; 90 m
4 (a) 6 m/s^2; 0; -3 m/s^2 (b) 300 m;
   600 m; 600 m   (c) 37.5 m/s
5 (a) 31.4 m/s   (b) 50.2 m
   (c) 22.2 m/s
6 (a) 6 m/s^2; 125 m   (b) 5 s; 100 m
7 (a) 103.8 N   (b) 5.19 m/s^2
   (c) 41.52 m; 21.96 m
8 (a) 392.4 m/s; 7848 m   (b) 10 010 N
10 (b) 157 N   (c) 882.9 N

Problems 6
3 331.3 m/s
4 5100 m/s; 3.4 m
5 (a) 200 kHz   (b) 10 cm

Problems 7
1 (a) 30 000 J   (b) 34 531 J
   (c) $2.4 \times 10^6 \text{ J}$
2 20 000 J; 6800 J
4 (a) 0.82 or 82 %   (b) 100 000 J
5 90 J

Problems 8
1 (a) 423 K; 234 K; 1273 K
   (b) $-223 \,°\text{C}$; $0\,°\text{C}$; $1227\,°\text{C}$
2 $59.2 \times 10^6 \text{ J}$
3 $509.5\,°\text{C}$
5 377 s or 6 min 17 s
6 4.9976 m; 4.9988 m; 5.0012 m;
   $1.92 \times 10^5 \text{ N}$

Problems 9
1 (a) $12\,\mu\text{V}$; 5 MΩ; 7 mA
2 (a) 4 A; 60 V   (b) 30 V; $R = 10\,\Omega$;
   $R_2 = 15\,\Omega$
3 (b) 1 A; 30 V   (c) 4.5 A; 1.5 A
4 2 mA; 100 V; 40 V; 60 V; 2.2 mA; 1.82 mA
5 (a) 2.5 A; 7.5 A; 10 A   (b) 20 Ω; 5 A
6 10 Ω; 2 A; 1 A; 0.67 A; 0.33 A
7 (a) 3 A; 1 A; 2 A   (b) 1.5 A; 1.25 A;
   0.25 A   (c) 2.25 A; 1.125 A; 0.225 A
8 (a) 100 Ω   (b) (i) 8 Ω/m (ii) 2 Ω/m
9 100 Ω
10 8.86 Ω; 8.20 Ω; 9.52 Ω
11 $0.004\,°\text{C}^{-1}$; 118 Ω
12 (a) 200 Ω; 232.6 Ω
   (b) $0.008\,°\text{C}^{-1}$ (c) 184 Ω; 280 Ω
14 $\alpha = 0.0047\,°\text{C}^{-1}$; $R_0 = 99.3\,\Omega$
15 (i) 374 mA   (ii) $\approx 400 \text{ mA}$

Problems 10
1 (a) 19.2 Ω   (b) 43.2 MJ   (c) 36 p
2 20 W; 28.8 W; 19.2 W; 68 W
3 3.48 V; 139.2 W
4 10 Ω; 1000 W
5 (a) 500 mA   (b) 250 mA
   (c) 8.33 A

Problems 11
5 10 V; 0.8 Ω
6 1.5 V; 2 Ω

Index